C·O·S·M·I·C O·D·Y·S·S·E·Y

How Intrepid Astronomers at Palomar Observatory Changed Our View of the Universe

Linda Schweizer

The MIT Press
Cambridge, Massachusetts
London, England

© 2020 Linda Schweizer

All rights reserved. No part of this book may be reproduced in any form by any electronic or mechanical means (including photocopying, recording, or information storage and retrieval) without permission in writing from the publisher.

Cover design, design, and composition by Briana Schweizer

This book was set in Garamond Premier Pro and Century Gothic in San Francisco, California.

Printed and bound in the United States of America.

Library of Congress Cataloging-in-Publication Data

Name: Schweizer, Linda Younker, author.

Title: Cosmic Odyssey : How intrepid astronomers at Palomar Observatory changed our view of the universe / Linda Schweizer.

Description: Cambridge, Massachusetts : The MIT Press, [2020] | Includes bibliographical references and index.

Identifiers: LCCN 2019057055 | ISBN 9780262044295 (hardcover)

Subjects: LCSH: Palomar Observatory--History. | Astrophysics--Research--California--History--20th century.

Classification: LCC QB461 .S3152 2020 | DDC 522/.1979498--dc23

LC record available at https://lccn.loc.gov/2019057055

10 9 8 7 6 5 4 3 2 1

To curious souls everywhere

Contents

FOREWORD BY DAVA SOBEL, page ix
PREFACE, page xi

· 1 · THE PROMISE, page 1
Architect of an Astrophysical Revolution | The "Big Eye" | Spherical Genius
The Palomar Sky Survey | Widening Horizons

· 2 · PLUMBING THE DEPTHS OF THE UNIVERSE, page 19
Blinking Beacons | The Swope Slope | A Glimmer of Things to Come
The Universe Throws Down the Gauntlet | The Whistleblower
"Stupendous Eruptive Phenomena" | Cinderella in the Subbasement

· 3 · UNRAVELING THE MYSTERIES OF STELLAR EVOLUTION, page 45
The Breakthrough | Stellar Radioactivity, Storms, Flares, and Winds
Nucleosynthesis Enshrined | Californium and the Bikini Test
Heavy Metal | Necropolis | Testing Einstein's Theory

· 4 · MILKY WAY ARCHAEOLOGY, page 69
Shooting Down the Dogma | Canaries in a Coal Mine
The Cannibalistic Milky Way | The Spaghetti Factory

· 5 · GALACTIC VIOLENCE: COLLISIONS AND MERGERS, page 85
Cosmic Changelings | A Blind Spot | Nuggets from a Survey
A Field Guide to Oddball Galaxies | The Whirlpool Galaxy and Its Companion
The Antennae Galaxies (NGC 4038/4039) | Ring Galaxies: Ghostly Apparitions
Forging a New Paradigm

· 6 · QUASARS: WOLVES IN SHEEP'S CLOTHING, page 111
From Local Hiss to Distant Roar | Radio Stars? | Energy Crisis
Agony and Ecstasy | A New Constituent of the Universe
Fuzzy Wuzzy Was a Quasar | Quasars Throughout the Universe
The Rise and Decline of Quasars

· 7 · PIERCING THE GALACTIC "SMOG", page 139
Far Sighted | Protostars: Galactic Vacuum Cleaners
Evolved Stars: Galactic Smokestacks | Galactic Center, Where Art Thou?
From Oddballs to LIRGs | Ultraluminous Infrared Galaxies: Celebrity Rock Stars

· 8 · STARBURSTS, SUPERWINDS, AND SUPERMASSIVE BLACK HOLES, page 157
Flashing Galaxies and the Provenance of Helium | Galactic Bubbles
Galactic Winds | Star Formation in the Middle of Nowhere
Maelstrom in the Nucleus | Black Holes: Central Monsters
The Key to the Vault

· 9 · PROBING THE GASEOUS UNIVERSE, page 181
A Spike of Light | Perfect Instrument Meets Perfect Telescope
Quasar Absorption Lines: A Biometric Passport
Gamma-Ray Bursts: High-Beam Dazzlers

· 10 · FROM GHOSTS TO GALAXIES: THE EMERGENCE OF STRUCTURE, page 195
Distant Shadows Red and Blue | E+A Galaxies | Ring of Fire
The Far Side of the Universe | Galaxy Clustering and Lyman-α Blobs | Violent Winds

· 11 · SOLAR SYSTEM SHUFFLE, page 213
The Buccaneers | Jupiter's Hotspots | From Frying Pan to Fire | Lord of the Planets
Incoming! | Bolide Flashes and Splashes | Solar System Refrigerators
Fragments from the Oort Cloud | Tallying Pieces of the Puzzle
Anatomy of a Murder: The Demise of Pluto as a Planet | Seeking Planet X: Hindsight
Seeking Planet Nine

· 12 · ASTRONOMICAL EXOTICA: NEW FRONTIERS, page 239
Runts of Star Formation | The Right Stuff
Hitting Pay Dirt with a Common Little M Star | Floating Dirt, Iron Rain, Magenta
Skies, and Violent Storms | In Search of Pale Blue Dots | Pathfinder
Transients, Superluminous Supernovae, and the Future of Observational Astronomy
The Transient Universe | The Fountainhead

ACKNOWLEDGMENTS, page 271
NOTES, page 277
INDEX, page 293

Foreword

The most foolish question I ever asked an astronomer concerned a mammoth construction project that had already eaten up years of the man's time. *Won't you be relieved*, I ventured, *when you can ditch the hard hat and get back to the real business of observing?*

The scientist made no effort to hide his annoyance at my naivete. Building this instrument, he informed me coldly, was the single most important work of his life. He doubted whether anything he might discover with its aid would surpass the value of having guided the great light bucket from idea to realization.

I imagine George Ellery Hale might have told me much the same thing, had I been around to interrupt his activities. Ever hungry for more light, Hale managed to build the world's largest telescope four times in the course of his career. He placed his first record-breaker, the 40-inch refractor of the Yerkes Observatory, near his native city of Chicago. That telescope (still the giant of its kind) was idled by the miserable local weather as many as two hundred nights per year. Hale turned west and lit out for California. He capped Mount Wilson with huge domes to house a 60-inch and later a 100-inch reflector. For his final triumph, he envisioned and initiated the fabrication of the 200-inch Hale telescope here on Palomar Mountain, but he did not live to see it see first light.

The magnificent instrument drew astronomers to Palomar. It even drew a few scientists from fields outside astronomy—for the promise and possibility of turning the Big Eye, as the Hale telescope came to be known, to some previously unexpected purpose. For although the 200-inch was attuned only to the wavelengths of visible light, it soon became an essential ally in the burgeoning fields of radio and infrared astronomy, and went on to help identify and characterize the sources of celestial outbursts of gamma rays and X-rays.

Findings made at Palomar repeatedly expanded the size of the known universe and altered prevailing views of its nature. Long before the Hubble Space Telescope endeared itself to earthlings as "the people's telescope," the Big Eye on Palomar Mountain funneled the cosmos into the popular imagination. What had once been seen as a realm of slowly evolving majesty, with isolated galaxies stealing ever farther away from us and from each other, was here revealed to be replete with violence: despite the vastness of space, galaxies could joust like enemy armies and plunder one another's stars.

Observing in the early days of the 200-inch in the 1940s and 1950s meant climbing into its prime-focus cage and guiding the telescope through the dark as though it were a balky animal. Yes, the air was frigid, the quarters cramped, the stress considerable, but many who braved those nights recall them as joyrides into the universe. As one veteran observer interviewed for this book recounted, "I was one hundred some-odd feet above the ground, and there was nobody else in the world. It was just me and the sky."

To say that a telescope functions as an extension of the astronomer's senses is to overlook the deeper attachment that can form between them, as together they peel back the layers of interference between curiosity and knowledge.

Some years ago I attended a weeklong symposium of the International Astronomical Union, held in Padua. The organizers had arranged for one of Galileo's original handmade telescopes to be put on display in a Plexiglas case in a central meeting area. The sight of it electrified the participants, who represented observatories and space agencies from all over the world. At any moment in the proceedings a cluster of them could be found standing transfixed around Galileo's telescope. Compared to the equipment they were accustomed to handling, this cardboard and leather tube appeared as ancient and rudimentary as a flint axe, but they regarded it—you could almost say they venerated it—as the fountainhead. Many of them carried childhood memories of getting hooked on astronomy with a starter scope more or less the same overall size and shape.

Today, most operations at Palomar proceed via robot control, and no one stands in the domes when observations are under way. The several instruments have been augmented through the decades with new detectors to extend their range, and adapted with optics to sharpen their view through Earth's atmosphere. As the newfound ability to collect gravitational waves from stellar collisions widens the scope of astronomical research beyond the confines of the electromagnetic spectrum, this observatory remains vital, even beloved.

Linda Schweizer's informed account of the science conducted here is an astronomer's answer to *The Thousand and One Nights*. It speaks of geeky midnight assignations, of dogged devotion to data gathering, of theories tested and patience tried though rarely exhausted. Her separate stories flit from light to dark matter, from quasars to extrasolar planets, starburst supernovae, and other exotica. The tales are peopled by a large cast of characters who collaborate and compete, working with or for or around one another in the unique environment of Palomar. One smitten observer—an astronomer-turned-astronaut—admits to pilfering a small piece of the place, which he carried with him into orbit for his first mission aboard the space shuttle.

Dava Sobel

Preface

The benefits of astronomy accrue to the human mind, as telescopes gather neither gold nor jewels except as people experience them in thought.

—Astronomy and Astrophysics Survey Committee,
chaired by John N. Bahcall, 1990

Many people dream of the stars. This book follows the astronomers who turn their dreams into hypotheses and discoveries on a cosmic scale. The "Big Eye"—as Palomar Observatory's largest telescope is affectionately known—played a significant role in those discoveries through much of the 20th century.

The casting of the Big Eye's 200-inch-diameter Pyrex mirror, the construction of its massive dome and horseshoe mount, and its first images of the universe generated as much public excitement as would the Apollo moon missions and the Hubble Space Telescope decades later.[i] Dedicated as the George Ellery Hale Telescope in 1948 after the leading American astrophysicist and astronomical entrepreneur, it was constructed with funds provided by The Rockefeller Foundation. It was the world's largest reflector for nearly 40 years and continues working at the forefront today. Just as Mount Wilson astronomer Edwin Hubble opened up the boundaries of the universe in the 1920s, so astronomers at Palomar wrote and rewrote the textbooks of astronomy in the mid to late 20th century. Through the Big Eye, the world encountered quasars and supermassive black holes, understood the chemistry that turns stardust into life, and pressed the limits of the known universe relentlessly outward.

The stories of Palomar's creation and its hero-astronomers such as George Ellery Hale and Edwin Hubble have been engagingly told in other works. The questions that drove me to write this book were about its cumulative contribution to science: What did this iconic observatory contribute to the trajectory of our understanding of the universe? And how did the discoveries happen and mature from theory to accepted paradigm? My own background as a PhD astronomer had taught me that such pathways are rarely simple, clear, or singular. I wanted to understand the rich scientific yield of more than 75,000 telescope-nights of observing distributed among Palomar's four main telescopes: the 200-inch Hale reflector, the 18-inch Schmidt camera (1936–mid 1990s), the 48-inch Schmidt camera (1948–current), and the 60-inch telescope (1970–current).

Preface

Cosmic Odyssey pulls back the curtain on how science is done by eccentric yet inspiring researchers—from drawing new insights out of raw images to the rivalries and collaborations that fueled ambitions and molded views. It tells the story of how different facets of the universe were explored, from the birth and evolution of stars to the discovery of quasars; from colliding galaxies to merging black holes; from the rubble of the solar system's formation to exoplanets circling other suns. How did astronomers open up new windows of the electromagnetic spectrum to unveil starbirth, the nucleus of our Galaxy, and quasars in the distant universe? This is not an exhaustive history, but one that follows the threads of significant discoveries from origin to culmination.

The twelve chapters illuminate the mechanics of science during nearly a century's work by the famous, the forgotten, and the lone maverick. Generations of men and women grappled with vast energies and mysterious processes far beyond human experience and imagination. Sometimes grudgingly, they were forced to abandon previous concepts to grasp extraordinary new wonders. The stories unfold here much as they did for the researchers, with gaps, diversions, dead ends, leaps, and suspense. Sometimes threads dip back 10, 20, or even 50 years or across continents to other teams and telescopes to retrieve a relevant clue. The book opens with the keystone to all the discoveries that follow: defining a "distance ladder" to calibrate the intrinsic properties of celestial objects, moving out from our solar system to the nearest stars and on to galaxies and clusters of galaxies and to the farthest reaches of the observable universe.

The narratives are anchored by the raw images and data plots that pioneering observers generated in long-gone eras, as well as by the technology and computer analysis at today's cutting edge. How did early explorers wring so much insight from a few grainy photographic spectra—a millionth of the data collected today? Astronomers have ever been challenged by technology, dim light, and new phenomena. Their thoughts and actions emerge for you, the reader, to share the *aha!* moment of understanding with them. What fascinates me is how the discrete insights acquired over decades by researchers in a global community cascade, collide, and—finally—coalesce.

Science is a creative and imaginative—even artistic—process. The stories of more than one hundred astronomers, cosmologists, physicists, planetary astronomers, and engineers who devoted their lives to studying all aspects of the universe provide long looks into the trenches where science is done. These individuals, some of whom were active in 1948, at the 200-inch telescope's first light, battled uncertainty and self-doubt, challenging technology and frigid cold. They were driven by their passion and, often, their ambition. Sometimes their discoveries were the result of luck or caprice. Although their goals are lofty and their dedication admirable, astronomers are not above racing their rivals to the finish line, ostracizing the underdog, or suffering from mental rigidity. United in their common quest, they also form decades-long partnerships and share data and hypotheses generously. When researchers tell me their stories, this becomes apparent: progress in science is rarely either well defined or directed, despite committees convened to hash out answers to the big questions.

Astronomers' stories portray the breadth of approaches to doing science. They also reveal the vulnerability, intense fervor, and dedication of the scientists who created entirely new fields in astronomy. My conversations with them yielded insider accounts of all-night rides in the cage atop the 200-inch telescope, of observers shivering in darkness amidst ice-cold metal instruments, of the thrill of epiphany and the thorn of uncertainty. One astronomer (Allan Sandage) compares Palomar to Xanadu and the pleasure domes of Kubla Khan, while another (Fritz Zwicky) labels his antagonists "spherical bastards." Yet another (Chip Arp), being strongly artistic, indulges his passion for cultivating orchids at home when he is not photographing peculiar-looking galaxies with the telescopes at Palomar. To one high-energy astrophysicist turned astronaut (John Grunsfeld), observing at Palomar was "a great romantic adventure" much like his operating a suite of telescopes aboard the space shuttle *Endeavor* three years later.[ii] No matter their stance, astronomers ultimately laid bare a universe of grueling complexity, profound mystery, and infinite beauty that no one could have foretold.

Imagine observing at Palomar: the whirring of the telescope and the rumble of the dome moving in the darkness, the oily smells, the "We're there!" of the telescope operator, looking up at the starry sky from the catwalk, and watching the marine layer creep up the mountain and stop just below the domes as if to insulate you on "Astronomy Island." This is the *Belle Époque* of cosmic exploration, and we're all participants.

During many stays at Palomar's Monastery, I had the privilege of joining astronomers and technicians at work in the 200-inch data room, chasing and recording objects ranging from ragtag newborn galaxies to exoplanets. During engineering breaks I crawled inside the cramped, oil-drenched arms of the Big Eye's massive horseshoe, where early infrared astronomers had long ago installed their cooling apparatus and detectors, and I peeked out from the prime-focus cage perched near the apex of the 135-foot dome. I swept my hand over the smooth circular rails that carry the thousand-ton dome as it rotates in lockstep with the sky's diurnal motion. I donned Tyvek gear to help the crew clean the mirror, then ogled in wonderment as a new coat of vaporized aluminum drifted onto its surface. I touched workhorse gears forged nearly 90 years ago, and admired the exquisite choreography of skilled mountain personnel maneuvering backhoes inside the dome as they swapped out

Preface

delicate car-sized instruments. Such extraordinary moments kept me inspired while toiling over these stories. They also reminded me that the scientific method is only part of discovery, just as it is only part of this book. Human cleverness, courage, and sometimes foibles are equally essential, and just as fascinating.

You, the reader, are now invited to travel on a wild ride of discovery with some of the greatest minds of astrophysics as they unlock the secrets of the universe from the top of Palomar Mountain. The story unfolds from the personal interviews and research materials of the protagonists, and as seen through their eyes, minds, and hearts.

Checking in at the Monastery. Photo by the author.

· 1 ·
The Promise

CHAPTER 1

The evolutionary history of galaxies is laid out in this deepest-ever image of a tiny patch of night sky. Except for a sprinkling of Milky Way foreground stars, each of the several thousand colorful smudges represents an entire galaxy—perhaps one of the first tiny blue seedlings that formed nearly 13.7 billion years ago, or the tortured remnant of a violent galaxy collision and merger, or a spiral, elliptical, or amorphous galaxy that we see closer by. The observable universe is estimated to contain more than 100 billion galaxies comparable to our own, and countless billions of fainter ones. The image portrays a patch of sky named the Hubble eXtreme Deep Field (XDF) in the constellation of Fornax. It was taken with the Hubble Space Telescope in ultraviolet to near-infrared light and covers about one ten-millionth of the entire sky. G. D. Illingworth et al., "The HST eXtreme Deep Field (XDF): Combining All ACS and WFC3/IR Data on the HUDF Region into the Deepest Field Ever," *Astrophysical Journal Supplement Series* 209 (2013). Credit: NASA; ESA; G. Illingworth, D. Magee, and P. Oesch, University of California, Santa Cruz; R. Bouwens, Leiden University; and the HUDF09 Team.

Adrift in a cosmos whose shores he cannot even imagine, man spends his energies in fighting with his fellow man over issues which a single look through this telescope would show to be utterly inconsequential.

—Raymond B. Fosdick,
President of The Rockefeller Foundation,
at the dedication of the 200-inch Hale telescope on June 3, 1948

In all of human history, few epochs have seen our view of the universe change as dramatically as have the past 100 years. Building on significant advances made by the older Mount Wilson Observatory, Palomar Observatory played a key role in this scientific revolution. Many threads of astronomical discovery trace back to the first light of the 200-inch telescope in the late 1940s, to a handful of pioneering men and women developing the foundations of modern astrophysics. Their paths of research met twists and turns, dead ends and disappointments, and occasional triumphs—realities of scientific practice. Thanks to generations of researchers across the world persevering, humankind's view of the universe—cobbled together through millennia—was transformed literally and figuratively in the cosmic blink of an eye.

CHAPTER 1

ARCHITECT OF AN ASTROPHYSICAL REVOLUTION

The transformation was unexpected and dazzling. It accelerated dramatically toward the end of the 19th century, as new tools developed in experimental physics and technology were honed for astronomical applications. One of the foremost leaders of the scientific and technological revolution in the United States was an enterprising young astrophysicist named George Ellery Hale, who dedicated his life to building telescopes powerful enough to decipher the light of the sun and stars. Born in Chicago, Illinois in 1868—just after the Civil War—Hale came of age during a period of growing synergy between astronomy and physics. Already as a student he had built an observatory and designed a spectrograph. After studying physics, engineering, and chemistry at the Massachusetts Institute of Technology, he began a distinguished career of research in solar physics while also emerging as a bold and ambitious scientific entrepreneur.

Traditionally, astronomers had counted, classified, and measured the positions and dimensions of celestial objects, with little interpretation or theory other than classical celestial mechanics. In contrast, these same objects beckoned to others as tools for studying the extremes of temperature, density, and mass unattainable in Earthly laboratories. Hale was keen to move past solely descriptive methods and calculate the heat generated by celestial bodies, to investigate their chemical compositions, and to understand how they might evolve.

Grasping the why and how of stellar physics would depend on more than counting and classifying. It would require filling large gaps in existing laboratory data on the temperatures, densities, and ionization stages of hot gases. Having learned early on that high-quality spectra of celestial objects were essential, Hale set out to materially improve both observing techniques and laboratory methods. Abandoning the old notion of an astronomical "observatory," he launched a prototype of his new vision: a combination of telescope and physical laboratory prepared for whatever the future might bring. Thus, a machine shop for fabricating precision equipment and a variety of state-of-the-art instruments borrowed from physics—such as spectrographs, custom gratings, thermocouples, and photoelectric cells—became integral parts of each of his successive observatories. On-site recording, measuring, and interpretation of the spectra of celestial objects became de rigueur.

THE PROMISE

George Ellery Hale at the spectrograph of the 60-foot solar tower telescope, Mount Wilson Observatory, circa 1908. Credit: Observatories of the Carnegie Institution for Science Collection at the Huntington Library, San Marino, California.

Hale masterminded the design and construction of four consecutive world's-largest telescopes, all dedicated between 1897 and 1948. The first, a refractor with an objective lens of 40-inch diameter, was erected at Yerkes Observatory in the plains of Wisconsin. With its long winters and poor seeing conditions, this location was an unwise choice. The inexperienced young astrophysicist understood his mistake from a comparison between his Yerkes telescope and Lick Observatory's 36-inch refractor set on a mountaintop in California; the latter yielded far superior observations. So, given the opportunity, Hale soon boarded a train and carried his dream from the smoke- and moisture-laden atmosphere of the Great Lakes to Southern California's clear skies

Chapter 1

and lofty mountains. He founded Mount Wilson Observatory in the mountains near Pasadena, California and had two reflectors built, one with a 60-inch-diameter mirror dedicated in 1908, a second with a 100-inch-diameter mirror—named the Hooker telescope—dedicated in 1917. Concomitant with that effort, auxiliary instruments were designed to analyze and study the physical conditions in stars. So equipped, the two telescopes harbored at Mount Wilson would soon challenge humanity's egocentric view of the universe.

The world was stunned in 1924 when observations by Mount Wilson astronomer Edwin Hubble, pushing the 100-inch Hooker telescope with all his might, bore fruit. Setting his sights on the Andromeda Nebula (as it was then known), the largest of many spiral-shaped nebulae dotting the night sky, Hubble established that its relatively modest apparent size and faintness were an illusion. Instead of being some small gaseous floater inside our Milky Way galaxy, the Andromeda Nebula turned out to be an enormous, full-fledged luminous galaxy about 2 million light-years away. It and its many sister nebulae in the sky were each comparable to our own Galaxy, "island universes" so to speak.

It was a dramatic adjustment for astronomers—as well as the public—to accept that our Galaxy, until then thought to be the entire universe, was instead just one of many thousands of similar galaxies in a millionfold-larger universe. In some sense, Hubble's landmark discovery rounded out the scientific revolution begun in the first decades of the 16th century when Renaissance astronomer and mathematician Nicolaus Copernicus circulated his belief in a heliocentric rather than geocentric universe. This finding was followed in 1609 by Galileo Galilei's discovery, with the newly invented telescope, of four moons orbiting Jupiter. Between 1915 and 1917, Mount Wilson astronomer Harlow Shapley resized the Milky Way galaxy by a factor of ten, mapped its shape, and learned that the sun lay in its outskirts, not at its center. So, when in 1924 Hubble announced his discovery, it was a coup de grâce, confirming that Earth is not at the center of our solar system, nor of our Galaxy, nor of the universe.

There was more to come. A few years later, as Hubble plotted the line-of-sight velocity of several galaxies against their distance, he noticed a striking correlation: the farther a galaxy from our Milky Way, the faster it recedes from us. In other words, the fabric of the universe is not static, as Einstein had assumed in his theory of general relativity, but instead expands. Hubble's discoveries prompted Einstein to visit Mount Wilson to see the 100-inch telescope and the new data for himself. He then realized that his "cosmological constant"—intended to force static models from his beautiful theory—was an ugly add-on, and he removed it. In a twist of fate, the cosmological constant would be repurposed seven decades later to characterize the mysterious dark energy responsible for accelerating the universe's expansion.

Even though Mount Wilson astronomers had made significant progress, galaxies themselves remained enigmatic, and little was known about the contents of the universe beyond a few dozen of the nearest galaxies. Although there was a rich store of diverse objects scattered across the sky at which to point telescopes, no instruments yet existed that could undertake a detailed inventory. Hale's questions deepened: How did the universe originate? How were the chemical elements in it created? How did stars and galaxies form, and how do they evolve? It wasn't long before he realized that the Hooker telescope could not see far enough into the universe to address these questions. It could not resolve most individual stars even in the Andromeda galaxy, one of our nearest neighbors. As the universe itself appeared to stretch to greater and greater depths, the Hooker's mirror could not keep up. Hubble—one of the most gifted astronomers on the mountain—was barely grazing the boundaries of the universe.

The "Big Eye"

It became clear to Hale that his succession of largest telescopes in the world could not be allowed to end with the 100-inch Hooker. In 1928, although approaching 60 years in age and plagued with fragile health, nervous exhaustion, severe headaches, and even hallucinations, his passion for science would not let him retire. Fortunately, Francis Pease—a determined Mount Wilson optomechanical engineer and astronomer who also dreamed of a larger telescope—had several years earlier produced drawings and a scale model for a 300-inch telescope (later downsized to 200 inches). So Hale, now director emeritus of the Observatory, launched a new project to build a cosmology machine powerful enough to answer his questions.

The years preceding the Great Depression of 1929 saw the accumulation of hefty private fortunes, and Hale was adept at hobnobbing with the wealthy. As he sought a benefactor for his latest dream, Wickliffe Rose, president of The Rockefeller Foundation's International Education Board, happened to be scouting for "brains to fund."[1] An intellectual giant himself, Rose had dedicated his life to the search for truth based on scientific inquiry. Hale hoped to engage Rose's interest by writing a compelling article for the April 1928 issue of *Harper's Bazaar* magazine. Titled "The Possibilities of Large Telescopes," his article opened with the poetic: "Like buried treasures, the outposts of the universe have beckoned to the adventurous from immemorial times." Hale then laid out profound scientific questions that could be answered only with a new giant telescope, further bolstered by recent advances in physics and chemistry.

CHAPTER 1

Hale's strategy worked. To Rose, the giant telescope symbolized a journey billions of light-years from Earth and to distant frontiers of the human mind. Within a few months, the Rockefeller Board had pledged $6,000,000—the largest-ever grant for a scientific project at the time—toward what it considered the most magnificent venture of its entire natural-science program. Due to a variety of circumstances, Rose awarded the grant to the California Institute of Technology, with its graduate programs, rather than to Hale's home base at the Mount Wilson Observatory, a pure research department funded by rival Andrew Carnegie's foundation. However, at the dedication of the Hale telescope in 1948, Caltech's president Lee A. DuBridge acknowledged: "Though this observatory was built officially by the California Institute of Technology it was conceived, planned and largely executed by the staff of the Mount Wilson Observatory."[2]

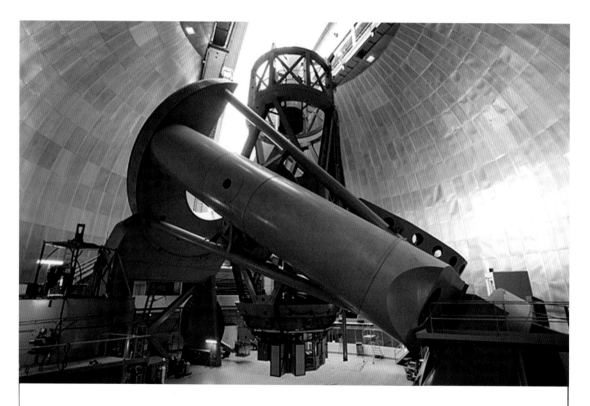

A recent image of the 200-inch Hale telescope with the prime-focus cage visible at the top of the support tube, the axis of the horseshoe pointing to the celestial north pole, the west arm of the mount visible in front, and the Cassegrain cage and electronics racks bolted in place under the mirror. Image courtesy of Rebecca Oppenheimer.

Although Caltech had noteworthy graduate programs in other sciences, it had no astronomy department yet and no experience in the design and construction of telescopes. So the Rockefeller grant bore the condition that the astronomers and engineers of the Mount Wilson Observatory would play an active role in the design and construction of the new telescope and observatory. Hale estimated that it would take four to six years to design, build, and install the telescope, auxiliary equipment, domes, and housing. But the engineering concepts, materials, and construction technology for casting the mirror and building its precision mount had not yet been developed. World War II and the realities of such an immense project intervened, and the project took 20 years to complete. Suffering from complete mental exhaustion, in 1936 Hale became too ill to lead the project. The Rockefeller Foundation's director, mathematical physicist Max Mason, left his post to head the Palomar Observatory Council from 1936 until the project's successful completion in 1948. Sadly, neither Rose nor Hale survived to witness the dedication of their engineering triumph.

As the future giant telescope promised to see far enough into the universe to permit the testing of his theory of general relativity, theoretician Albert Einstein took an interest. His equations had predicted that the curvature of space depends on the mass of the entire universe, which could only be discerned from the motions of objects very far from our Galaxy. On April 30, 1937, Einstein, along with 200 scientific and industrial leaders, celebrated the completion of the massive truss tube for the 200-inch telescope mount—"wide enough to engulf a 2-story bungalow"—before it was shipped to Palomar. In a photograph, Einstein is shown examining a 1:32-scale celluloid model, built by Westinghouse Electric and Manufacturing Company to test points of stress and the feasibility of their design. Image courtesy of Westinghouse Electric and Manufacturing Company.

Chapter 1

Today it is hard to imagine how the engineers of the 1930s and 1940s managed to shape more than 500 tons of steel and glass into a precision instrument that—mated to its 1,000-ton rotating dome—seems so breathtakingly flawless. As with astronomical equipment built today, the engineering design goals for the 200-inch telescope were dictated by the extreme optical accuracy required by astronomers' work. But unlike today, there were no standard practices to follow, no prototypes, no coefficients of torsion, and no computer models. The designers' basic tools and techniques included slide rules, difficult hand calculations, ample pencil sketches, drafting, blueprints, and occasional sparks of brilliance—all abetted by years of engineering experience. The engineers tested their innovative designs by experimenting with functioning paper, wood, or metal scale models of the dome, telescope, and drive system.

Aware of how groundbreaking the giant telescope might be, workers struggled to cast 20 tons of Pyrex—a borosilicate glass with low thermal expansion—into a nearly 17-foot-diameter disk at Corning Glass Works. In 1936, millions of people swarmed railways and roads to watch the fragile disk cross the country through tunnels and over bridges from Corning's New York factory to Caltech's optical shop. Workers at General Electric Corporation and Westinghouse Corporation were tasked with cutting, welding, and bolting 500 tons of unwieldy metal into structures capable of supporting, protecting, and pointing the mirror with exquisite accuracy.

Meanwhile, workers on Palomar Mountain were constructing the enclosure for the new telescope, a cylindrical Art Deco building topped with a double-walled rotating steel hemisphere with a long slit and shutters. The dome soared 135 feet above ground, nearly to the height of the oculus of Rome's Pantheon. The telescope's pedestal was anchored 25 feet deep in solid granite at the mountain's 5,600-foot summit.

During the construction period, the public was captivated by news stories about Palomar's astronomers-in-waiting, the abundance of clever design solutions and engineering feats, and the harrowing transport of tons of glass, steel, and concrete to the top of Palomar Mountain. In addition, the name recognition of The Rockefeller Foundation and Mount Wilson Observatory, coupled with the unprecedented size of the grant, provided plenty of fuel for public interest. The gargantuan telescope was branded the "Big Eye" by the media, whose power to charm was just emerging. The project's appeal to these different audiences was strong enough to carry it through two decades of global depression and war. Among astronomers and physicists, the promise of the giant telescope generated feverish anticipation and planning for how best to explore the great mysteries of the universe that might now be in reach.

Spherical Genius

No one could have foretold the critical role that a Swiss-national physicist would play in shaping not only the legacy of Palomar Observatory, but the entire astronomical research enterprise well into the 21st century. In 1925, Fritz Zwicky, having earned a doctorate in X-ray crystallography from the Federal Institute of Technology in Zurich, joined Caltech's physics department to commence a two-year fellowship from the International Education Board of The Rockefeller Foundation. Initially, he planned to continue his research in solid-state physics and crystallography. However, he soon became aware of the groundbreaking work in astrophysics and cosmology being done at nearby Mount Wilson Observatory, and of the plans for a telescope that would see twice as deep into the universe as the Hooker telescope. These events enticed him—and eventually a growing number of other physicists—to switch his allegiance from physics to astrophysics. Zwicky learned the elements of astronomy from Mount Wilson staff members, especially Walter Baade, a recent transplant from Germany. Besides classical topics such as the spectral characteristics of stars and galaxies, Baade drew Zwicky into the curious world of exploding stars—later dubbed "supernovae"—that for a brief time flare as brightly as an entire galaxy.

As the chance of catching a supernova *in flagrante* in any one galaxy is minuscule, large numbers of galaxies must be photographed frequently to discover such flare-ups. With their small fields of view, the 60-inch and 100-inch telescopes on Mount Wilson could photograph only a few galaxies at a time, insufficient for recording rare and unpredictable supernovae. Zwicky realized that even the 200-inch telescope—though it could record an abundance of more distant galaxies—would have a keyhole-sized field of view only 9 arcminutes in diameter, less than one-third the apparent diameter of the full moon. With the entire sky measuring 41,253 square degrees, of which three-quarters are visible in the course of the year from Palomar, the 200-inch would have to photograph the sky every night for an entire human lifetime to image even a small fraction of it. How would astronomers know where interesting objects were to be found? Zwicky wanted to survey and catalog the entire contents of the universe. The task would never be achieved with either existing or planned large telescopes.

Fortuitously, in 1929 a genial, whiskey-loving, one-handed Estonian-born optician named Bernhard Voldemar Schmidt had invented the perfect tool. While working at the Hamburg Observatory in Germany, he found a way to correct an optical deficiency of fast wide-field mirrors (mirrors with a short focal length relative to their diameter) by inserting a special correcting glass plate into the light beam. He built a

Chapter 1

prototype camera which not only produced wide-angle images, but images that were sharp and free of distortions out to the very edge of the field of view. Baade, then at Hamburg Observatory, was a close friend of Schmidt. When Baade joined the staff at Mount Wilson, he introduced Schmidt's revolutionary optical design to Zwicky.

Optician Bernhard Voldemar Schmidt is shown testing an instrument at the Hamburg Observatory. His invention of a wide-field camera revolutionized astronomy by enabling observers to survey the entire sky for faint galaxies and nebulosities. The Schmidt camera's combination of mirror and lens is widely used today not only in astronomy but also in aerial survey cameras, television cameras, and movie theater and home projectors. Credit: Hamburger Sternwarte.

So it is that after the location of the 200-inch telescope had been chosen and construction was about to begin, a prototype survey telescope based on Schmidt's invention was suddenly and belatedly inscribed into the project blueprint. Zwicky was put in charge of his pet project, which progressed quickly. The Palomar 18-inch Schmidt camera, which Baade, out of deep gratitude to his optician friend, would always refer to as "the Schmidt," was dedicated in 1936. This late addition would turn out to contribute immensely to the visualization and mapping of the universe by the global astrophysics community. But for the following 12 years the Schmidt camera was Palomar's sole sentinel on the universe, and Zwicky was its master.

Swiss-born astrophysicist Fritz Zwicky introduced the idea of surveying the sky in order to explore and inventory the contents of the universe. He is shown observing with the new Palomar 18-inch Schmidt camera around 1936. Since the Schmidt was designed solely for imaging (and not spectroscopy), it is, in effect, a wide-field telescopic camera. For this reason, the designations "Schmidt camera" and "Schmidt telescope" are often used interchangeably. Courtesy of the Archives, California Institute of Technology.

A survey camera as fast as the Schmidt (with focal ratio f/2) was unique. It could photograph objects as faint as the sky background 25 to 50 times faster than could conventional telescopes (f/10 to f/15), and twice as fast as the 200-inch telescope (f/3.67). In its first five months of operation, Zwicky photographed 100 fields that spanned one-sixth of the entire sky visible from Palomar. On the Schmidt's circular-cut films, each covering an area of the sky 8.5 degrees in diameter (17 times the apparent diameter of the full moon), he discovered wisps of luminous material that stretched like taffy around and between galaxies, intriguing rare galaxies with streamers, and even rarer galaxies

Chapter 1

that appeared like rings of bright stars. These kinds of galactic appendages and structures had never been seen before. Sometimes one film recorded as many galaxy clusters as the total seen up to that time by all previous telescopes. When Hubble expressed doubts that Zwicky had actually found a major cluster of galaxies in Pisces, the pugnacious astrophysicist—famous for labeling his antagonists "spherical bastards"—shot back that the 100-inch Hooker telescope could not cover even a tenth of such a cluster, while the 18-inch Schmidt could cover the whole. Zwicky expressed his hope that the value of powerful wide-angle cameras would now be apparent.

The Palomar Sky Survey

And it was. Zwicky's survey of the sky was so fruitful that it became a priority for the Observatory to build a second, larger version of a Schmidt camera that would be a better match for the "Big Eye" still under construction. With an additional grant of half a million dollars from The Rockefeller Foundation, the 48-inch Schmidt camera (or "telescope") was finished in 1948 in time for the 200-inch telescope's dedication on June 3. Within a year, the National Geographic Society—known for exploring the Earth—was invited to explore the sky and universe as never before by participating in a photographic survey of the entire sky visible from Palomar.

Thus, the 48-inch Schmidt spent its first seven years photographing 900 square fields, each six degrees on a side, and each consecutively with a blue- and a red-sensitive emulsion. The 14-inch-by-14-inch photographic plates, custom-made by Eastman Kodak, recorded objects one millionth the brightness of the faintest star visible by naked eye on a dark night. Hundreds of millions of never-before-seen objects showed up, such as individual and clustered stars and galaxies, nebulae, comets, and asteroids. Three-quarters of the entire sky was mapped by the Palomar 48-inch Schmidt, and the rest was later mapped with similar Schmidt cameras at observatories in Chile and Australia. The National Geographic Society–Palomar Observatory Sky Survey (known by its abbreviation NGS-POSS, or simply POSS) was a smashing success: the 48-inch Schmidt photographs could now show the Big Eye where to look deep into the universe.

The field of view of the Palomar 48-inch Schmidt camera (overall image) as compared to that of the 200-inch Hale telescope (small rectangle at the center of the field). Both telescopes were trained on Messier 87, the second-brightest galaxy in the Virgo Cluster of galaxies. The wide-angle Schmidt camera covers 6.4° × 6.4° on the sky (corresponding to 13 times the moon's apparent diameter) and is suitable for surveys. The 200-inch telescope zooms in to 0.42° × 0.33° (less than the moon's apparent diameter on a side) to capture fainter, finer detail (inset, lower right) at a scale six times larger than those taken with the Schmidt. Credit: NASA/B. Schweizer.

As the stories of scientific discovery unfold in the following chapters, it will become clear that the two Schmidt cameras and the NGS-POSS played as large a role as the 200-inch telescope in laying the foundations of modern astrophysics. Besides being a gold mine of objects to study, the collection of photographic plates provided an optical basis for locating and identifying strange sources observed at other wavelengths. While exclusive access to the 200-inch telescope ensured that Caltech and Mount Wilson astronomers would dominate cosmology and astrophysics for several decades, the generous distribution of copies of the survey plates to other astronomical observatories and research centers invited astronomers and graduate students worldwide to do cutting-edge research of their own. As an example, Russian astronomer Boris Vorontsov-Velyaminov was quick to study the newly released survey prints and published the first illustrated catalog of interacting galaxies before the California locals found time to do so.

Within a few years, Mount Wilson and Palomar astronomers used more accurate methods to show that the distance to the Andromeda galaxy—and to all galaxies in the universe—was twice what Edwin Hubble had claimed. Doubling the apparent size of the universe began to upset "business as usual." With sharper insight, the inferred size of the entire observable universe soon ballooned another eightfold, and it turned out that the famous discoveries made at Mount Wilson were based on a tiny fraction of the universe now available for exploration from Palomar. Yet even wilder surprises lay ahead.

WIDENING HORIZONS

For millennia we humans had viewed the universe only in the narrow wavelength band visible to our eyes. Even the introduction of photography did not broaden that band much. Then, with the end of World War II, new technology designed to deliver victory in the war was deployed to expand the range of wavelengths available to study the sky. Just as the new telescopes of Palomar opened their shutters to look for faint optical sources, fledgling radio telescope operators in Australia and the United Kingdom began to discover some puzzlingly bright sources of radio energy. They asked Pasadena astronomers to search for optical counterparts with the 48-inch Schmidt and 200-inch Hale telescopes.

By 1963, some optical counterparts were found that appeared to be blue stars inside our own Galaxy. However, spectra taken of the "stars" showed that their appearance was deceptive. In reality, these strange pointlike sources lay billions of light-years from Earth and were disgorging titanic energies—trillions of stars' worth—from tiny solar-system-size volumes. Acting as powerful searchlights throughout the far reaches of the universe, the objects, soon dubbed "quasars," revealed the gaseous medium filling the universe. Understanding that a staggering amount of energy was being generated by a tiny source led to intoxicating ideas about supermassive black holes, relativistic gravitational collapse, and the omnipresence of energetic particles and explosive events. Suddenly, the universe had grown far more violent and complex than just a bunch of stars glittering peacefully in the sky.

The increase in complexity continued. The radio-astronomical revolution was about a decade old when some physicists began to pry open a second window of discovery in the infrared. Conservative astronomers, accustomed to studying the universe only with optically sensitive photographic plates, at first regarded the bulky infrared instruments as a nuisance. They changed their tune as they realized that the universe is laden with dust that blocks optical light, while allowing infrared photons to pass freely through. Newly equipped with state-of-the-art infrared detectors that had been skillfully "glad-handed" from the Naval Air Weapons Station at China Lake, California, the Mount Wilson telescopes and Palomar's Big Eye offered astronomers provocative glimpses of vast unexplored territories. They saw stellar nurseries deeply embedded in cocoons of native gas and dust, and aging stars polluting the Milky Way like smokestacks. There were bright gaping holes in Jupiter's cloudy atmosphere, and distant galaxies undergoing enormous bursts of star formation. Eventually, infrared astronomers peered through the Milky Way's smoggy disk into the heart of our own Galaxy to discover that a supermassive black hole lies hidden even there.

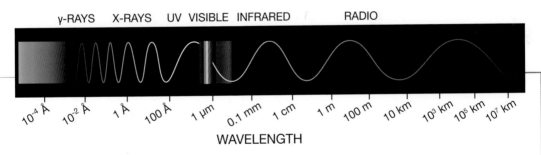

Early in the 20th century, most astronomical observations were conducted within a single octave of the electromagnetic spectrum, from 3,500 to 7,000 angstroms (0.35 to 0.7 microns, where a micron is one-millionth of a meter; an angstrom, abbreviated Å, is one ten-billionth of a meter). With the addition of radio and infrared detectors, and later X-ray and gamma-ray satellite observatories, the range in electromagnetic wavelengths accessible to early 21st-century astronomers has increased by more than a quintillion (10^{18})! The 18 orders of magnitude in wavelengths currently available to astrophysicists range from waves shorter than the diameter of a proton to waves the size of continents. However, wavelengths alone are not the whole story. Since Earth's atmosphere absorbs some radiation, windows of relatively high atmospheric transmission dictate the wavelengths at which astronomers can observe celestial sources relatively unimpeded. (Image courtesy R. Schweizer.)

Chapter 1

Metaphorically speaking, modern astronomers find themselves behind a closed door with a keyhole. Through this keyhole, be it a telescope on a mountaintop or one in space, photons and particles stream en masse from the entire expanse and history of the universe. Unlike other scientists able to perform experiments under controlled conditions in a laboratory, most astronomers cannot touch, experiment with, or modify their favorite objects of study. In spite of this handicap, during the past century we have constructed an impressively coherent picture of our universe. Approximately 13.7 billion years after the primordial explosion that marked the beginning of space and time, we discern a universe aglow with galaxies and gas strewn over a web of filaments surrounding enormous voids. We see planets dance around alien suns, galaxies spewing forth jets of plasma from their nuclei, and stars violently exploding to seed more structure and life itself. But even this proves only the tip of an iceberg.

There is ample evidence that most of the mass in the universe is dark, invisible, and, apparently, does not interact with ordinary matter or light. Although this "dark matter" rules the architecture of our universe, its nature remains unknown; perhaps it consists of an exotic particle not yet discovered. Additionally, there is recent evidence that the cosmic expansion is, shockingly, accelerating rather than slowing down, as had been expected from mutual gravitational attraction between its constituents. In response, cosmologists have proposed the existence of a pervading repulsive substance, dubbed "dark energy," that pushes everything in the universe apart, in a sense vanquishing gravity.

First Copernicus and Galileo, then Shapley and Hubble shattered our Earth-centric views. Now these recent advances in astrophysics have taught us that all we can directly observe with our telescopes amounts to a mere smidgeon of all the matter and energy that exists in the universe. The universe is overwhelmingly made of stuff of which we know little—and did not even suspect existed until a few decades ago.

Palomar designers' tools and materials. Photo by the author.

· 2 ·
Plumbing the Depths of the Universe

Chapter 2

Plumbing the Depths of the Universe

A fundamental goal of astronomy is not only to discover but also to *measure*. By measuring the pulsation periods and magnitudes of what he thought were Cepheid stars located in the globular clusters surrounding the sun, Harvard astronomer Harlow Shapley revamped our understanding of the shape, size, and configuration of the Milky Way galaxy. Similarly measuring pulsation periods of Cepheids in several nearby galaxies, Mount Wilson astronomer Edwin Hubble discovered that the universe is much larger than previously thought, and then went on to show that it is expanding. Although that revelation caught Albert Einstein's attention, the structure and fate of the universe lay frustratingly hidden in small-order effects at large distances. In his 1951 Penrose lecture that outlined work at the Hale telescope, Hubble expressed confidence that with sufficiently accurate data it would be possible to measure these effects. The fate of the universe could then be learned from the behavior of his law of redshifts—that the velocity of recession of a galaxy is proportional to its distance from Earth. A slight tilt up or down in an otherwise linear relation would divulge "whether the rate of expansion of the universe has been *speeding up* or slowing down during the immediate past."[3] It would take nearly half a century of precision measurements with ground- and space-based telescopes to find the shocking answer.

Sifting through her observations of thousands of variable stars in the Magellanic Clouds while working as one of Harvard Observatory's so-called "human computers," Henrietta Swan Leavitt discovered a near-perfect cosmic measuring tool. It is what astronomers have come to call a "standard candle," an object or class of objects of known luminosity. She announced in 1912 that the mean apparent magnitude (a measure of brightness) of a certain class of pulsating variable stars, named Cepheids, happens to depend on the period of pulsation only. (H. S. Leavitt and E. C. Pickering, "Periods of 25 Variable Stars in the Small Magellanic Cloud," Harvard College Observatory Circular 173 [1912]: 1–3.) Pictured here, the bright star RS Puppis, enshrouded in clouds of dust, is a Cepheid variable star. Because its outer layer is in disequilibrium, the star expands and contracts rhythmically to compensate for radiation that gets periodically trapped as it travels from the core to the photosphere. Every 40 days, as RS Puppis brightens by a factor of five, the reflection of its light in the surrounding dust clouds brightens correspondingly—but with a delay due to the light's travel time. From the delay between the brightening of the star and its "light echo," astronomers can calculate a precise distance to the star. Since Cepheids are massive stars and intrinsically bright—up to 100,000 times brighter than the sun—they serve as reliable standard candles for calibrating the cosmic distance ladder. Credit: NASA, ESA, and the Hubble Heritage Team (STScI/AURA)-Hubble/Europe Collaboration; acknowledgment: H. Bond (STScI and Penn State University).

Chapter 2

Blinking Beacons

The 200-inch Hale telescope was formally dedicated at Palomar Observatory on June 3, 1948. At first, with both Schmidt cameras and the 200-inch Hale telescope all working in tandem, Pasadena astronomers continued with "business as usual." They reexamined and refined programs that had been in place for a quarter of a century at Mount Wilson Observatory. However, cosmological studies that could only be advanced with the deeper vision of the Hale telescope ranked high on their agenda. These included the structure of the universe and the significance of its expansion, whether space is curved and bounded, or open, or flat. With more light now available for the spectroscopy of faint objects, astronomers hoped to crack the codes of the origin of the chemical elements and of stellar evolution. They also wanted to solve a disturbing astrophysical mystery that had become one of the main drivers for building the 200-inch telescope: namely, that the universe appeared to be paradoxically younger than its contents.

On the one hand, an estimate of Earth's age—taken at the time to be 2.8 billion years—was derived from the radioactive decay of uranium and thorium found in Earth's crust, a method deemed reliable. On the other hand, Hubble's discovery that the universe is expanding provided a way to estimate the latter's age. If galaxies are flying apart from each other, they must have originated from one small volume. Extrapolating their motions back in time to the beginning defines the age of the universe. The ratio between the galaxies' velocities of recession and their distances from us is the current rate of expansion of the universe. It is measured in units of (km/s)/megaparsec (where 1 parsec is 3.26 light-years and 1 megaparsec is 3.26 million light-years), and was until recently named the "Hubble constant," or H_o for short, and now the "Hubble-Lemaître constant." From it, Hubble derived an age of 1.8 billion years for the universe, clearly less than the age of Earth.

Some astronomers asked whether the anomalously low age of the universe might be due in part to Hubble's reliance on photographic photometry, a difficult, highly nonlinear technique for estimating magnitudes (a measure of a celestial body's brightness). Hubble openly admitted that his magnitudes were "suggestive and not definitive." Walter Baade, his younger Mount Wilson Observatory colleague, daringly called them "enthusiastic" and had serious doubts about the form of the Cepheid period-luminosity relation that formed the bedrock of Hubble's pioneering measurements.

Baade's doubts had developed during World War II when, as a German alien, he was confined to Pasadena and the Mount Wilson Observatory. Its 100-inch Hooker dome turned out to be a golden cage. Capitalizing on the blacked-out Los Angeles basin, he photographed the Andromeda galaxy (also known as Messier 31) to his heart's content. He discovered that the galaxy's young blue luminous giant stars and dust lanes appear preferentially in its disk and spiral arms, while myriads of old red stars tend to congregate in its central region and in its globular clusters.[4] He named the two physically distinct stellar groups "Population I" and "Population II," respectively. Although astronomers for 40 years, including Hubble, had lumped all Cepheids together in one period-luminosity relation, Baade wondered whether Cepheid variables from the two populations should be grouped this way. Admittedly, the variations in their brightness plotted against time—their light curves—appeared similar, but there was no a priori reason why stars with such different characteristics should be equally luminous.

This is where things stood in early fall of 1950, when Baade journeyed to Palomar Mountain to take his first photographs of M31 with the 200-inch telescope. The refrigerator in the dome was stocked with photographic plates coated with emulsions sensitive to blue light. Meticulous and technically gifted, Baade loved to observe and, like Edwin Hubble, always worked in coat and tie. He knew it would be difficult to detect small variations in brightness of individual stars so far away, especially in crowded fields of stars whose intrinsically steady light flickers with each random fluctuation in Earth's atmosphere. Thus, he worked only under the best sky conditions. It is said that he would rather let the world's largest telescope idle for an entire night than observe fuzzy or twinkling stars.

RR Lyrae variables are old, low-luminosity, low-mass Population II stars which served as standard candles in studies within the Milky Way galaxy. Baade had planned to compare the magnitudes of these variables side by side with Cepheid variables across the face of M31. If Hubble's estimate of the distance to M31 was correct, its RR Lyrae variables should have been visible just above the detection limit of Baade's plate. But it was clear from his first photograph that something was wrong. Instead of RR Lyrae, he saw only the brightest stars of Population II, which he knew were 1.5 magnitudes brighter than RR Lyrae variables. Since 1.5 magnitudes corresponds to a factor of four in brightness, this meant that the distance to M31 was twice what Hubble had determined.

Baade decided to derive the distance to M31 by another route, this time pairing Hubble's own Cepheid data with more accurate magnitude sequences newly obtained with photocells. This distance was close to what Hubble had claimed. Hence, there were now two determinations of the distance to M31, one from Baade's Population II stars and another from Hubble's Cepheids—and they differed by a factor of two.

Chapter 2

Although it was considered heretical at the time, Baade questioned Hubble's distance to M31. Having more confidence in his own data for Population II and RR Lyrae stars than in Hubble's Cepheid data, Baade concluded that M31 was twice as far away as Hubble had determined. Since M31 is a crucial rung in the cosmic distance ladder, doubling its distance meant doubling the distances to galaxies further away, out to the edge of the universe. It also correspondingly doubled the universe's age to 3.6 billion years, making it comfortably older than the then-current estimate of the geological age of Earth. In 1952, Baade announced his findings at the General Assembly of the International Astronomical Union in Rome. The story was picked up by popular magazines and newspapers and stirred the public's imagination.

THE SWOPE SLOPE

By 1952, the 200-inch telescope had brought Baade into a new realm of observing. The telescope detected objects about 1.5 magnitudes fainter than the 100-inch Hooker on Mount Wilson and could reach objects twice as far away. Baade hoped to directly compare standard candles in M31, such as young Population I Cepheids and old Population II Cepheids. Over the next few years, he took several hundred photographic plates of this galaxy, choosing the time intervals between plates so that Cepheid light curves could be constructed efficiently. Each visual-light plate took 2 to 2½ hours to expose, while blue plates took 30 minutes. For the first time in history, Baade hoped, the fundamental period-luminosity relations of the two kinds of Cepheid variables could be compared directly within the same galaxy. He expected that the Hale's deeper plates would help him establish the slope of Leavitt's period-luminosity relation by covering a wide range of luminosities, all the way down to the faintest, short-period Cepheids.

Knowing that his program would take many years to carry out, Baade warned that at times it might look as if the program had stalled, because everybody was in hot pursuit of new leads that opened up suddenly. But he assured his readers that it would be carried out, because "without a secure base we will go astray and finally become lost."[5]

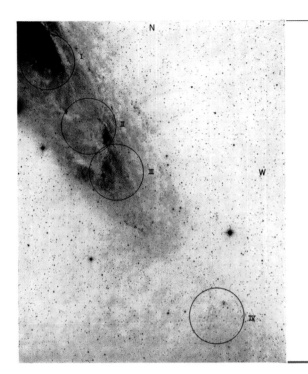

To conduct his search for Cepheid variables in M31, Baade concentrated on four circular star fields, which he marked on a 48-inch Schmidt plate. His aim was to find out empirically whether the period-luminosity relation of Cepheids changes with distance from the galactic center. So he chose fields at increasing distances from the center along the galaxy's major axis. Since absorption by dust can pose problems, Field IV was located far out in M31 to minimize the dust absorption and the crowding of stars. Plate 1 from W. Baade and H. H. Swope, "Variable Star Field 96′ South Preceding the Nucleus of the Andromeda Galaxy," *Astronomical Journal* 68 (1963): 435–469. © AAS. Reproduced with permission.

Swamped with photographic plates and working on several other interesting projects, Baade invited Henrietta Swope, an expert in photographic photometry of variable stars at Harvard College Observatory, to join him in Pasadena. Swope—whom he had met years earlier as she presented a research paper and again at an International Astronomical Union conference—eagerly accepted. Instead of working on relatively bright variable stars in our own Galaxy, as she had done with Harlow Shapley, she would work on much fainter variables recorded on Baade's 200-inch plates—and on galaxies fainter and farther away. Ira Bowen, director of Mount Wilson and Palomar Observatories, officially offered her a position and assigned her the job title "Computer." But Swope, normally quite shy, wrote back that she had never been called "Computer," only "Assistant" or "Staff Member." So Bowen changed her title to "Research Assistant."[6] As a woman Swope was not allowed to lodge at the Monasteries on Mount Wilson or Palomar. Instead, she had to work at her desk at the headquarters of the Mount Wilson and Palomar Observatories on Santa Barbara Street in Pasadena, turning Baade's enormous collection of plates into quantitative data. Down the hall from her office, she could hear Baade chatting excitedly with his colleagues, an excitement that she happily shared.

Chapter 2

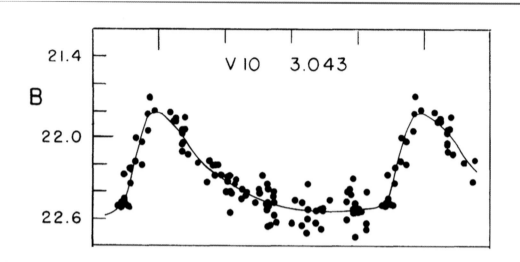

Cepheid variables exhibit characteristic sawtooth-shaped light curves, one of which is shown here as brightness plotted against time. The scatter of the data points is due to observational errors. The light curves of different Cepheids show minor variations due to intrinsic physical differences. As Baade's indefatigable research assistant, Henrietta Swope enjoyed the tedious, exacting work of measuring, plotting, and analyzing the magnitudes of the Cepheids that Baade had marked on his plates. Figure 8 from W. Baade and H. H. Swope, "Variable Star Field 96′ South Preceding the Nucleus of the Andromeda Galaxy," *Astronomical Journal* 68 (1963): 435–469. © AAS. Reproduced with permission.

Widely recognized for his observational skills and deep insight, Baade was the visionary of this project, although he had an aversion to publishing papers. In 1958 at the age of 65, he retired and left for an extended trip to Europe before publishing his magnum opus. In his absence, Swope continued to diligently quantify and transform the specks of light on Baade's plates into graphs, light curves, and tables. But Baade never returned to Santa Barbara Street. In 1960 he died of complications after surgery, leaving Swope to finish their project on her own. With a heroic effort, she pulled all of the data together and executed what Baade had hoped to determine for himself: finding separate period-luminosity relations for the two populations of Cepheids he had discovered.

Swope wrote papers in 1963 and 1965 on M31's variable stars, papers that Baade never saw, although she made him first author. She came up with a graph that would have pleased him tremendously; it confirmed the hunch that in 1952 had led him to propose the doubling of the distance to M31 at the IAU General Assembly. On this graph, the classical Population I Cepheids in M31 lie along a period-luminosity relation whose slope she took from works by Chip Arp (Baade's former student) and Robert Kraft on variables in the Small Magellanic Cloud and in the Milky Way, respectively. At the same time, her graph shows, with open squares, that the five Population II Cepheids, which lie way off this relation, are well fitted with the same relation shifted downward by two magnitudes.

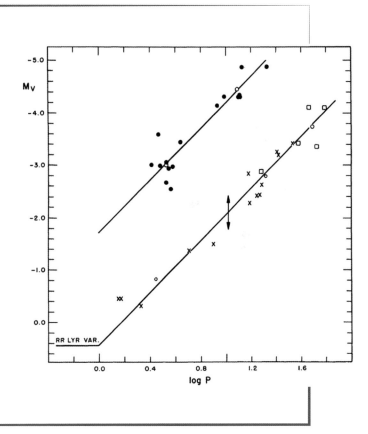

Two distinct period-luminosity relations determined by Henrietta Swope for the young, more luminous Cepheids of Population I (top curve) and old, less luminous Cepheids of Population II (lower curve). Interestingly, the relation for Population II Cepheids also fits the same kind of variables in Milky Way globular clusters (small crosses). This figure beautifully confirms what Walter Baade had concluded in 1952 from studying his first 200-inch plates of M31. Figure 15 from W. Baade and H. H. Swope, "Variable Star Field 96′ South Preceding the Nucleus of the Andromeda Galaxy," *Astronomical Journal* 68 (1963): 435–469. © AAS. Reproduced with permission.

Because Hubble had failed to recognize the distinct period-luminosity relations of these two Cepheid populations, he had underestimated the distance to M31 by a factor of more than two. It is the doubling of this distance, a crucial rung in the cosmic distance ladder, that had then doubled the size of the universe as announced by Baade.

Chapter 2

A Glimmer of Things to Come

In 1935, Hubble had initiated a program to resume where he had left off in 1929. His goal was to push the velocity-distance relation of galaxies to the furthest distances possible with the largest telescopes. Cooperation between astronomers from different institutions is standard practice today, as large surveys are routinely carried out with coordinated ground-based and space-based observatories. However, in the 1930s it was rare for astronomers from competing observatories to pool their data into one publication. In an unusual move, Hubble charmed his colleagues at two California institutions to join him in a collaboration that would last 20 years.

The work was meted out according to the observational limits of the largest instruments available at each observatory. At first, Milton Humason worked the 100-inch Hooker telescope on Mount Wilson to its limits, then he and Allan Sandage used the 200-inch Hale telescope at Palomar when it became available. Nicholas Mayall observed with the 36-inch Crossley reflector at Lick Observatory. The larger telescopes were tasked with gathering redshifts and apparent magnitudes for 600 faint distant galaxies, including members of 26 clusters, while the Crossley was trained on 320 nearby bright galaxies. More than one hundred galaxies in common were used to check for systematic errors.

When Hubble died in 1953 before his program was completed, Sandage was called in to analyze the data and decipher the workings of the universe. One of the most powerful observational probes at the time was the so-called velocity-distance relation for galaxies. The shape of this relation, which plots a galaxy's apparent magnitude (a measure of brightness) against its recession velocity (the velocity with which the galaxy moves away from our Milky Way galaxy) depends in a predictable way on the geometry and kinematics of the universe. After laying out the various corrections to the raw data, Sandage plotted the velocity-distance relation, since renamed the Hubble diagram, for the entire catalog. Although the shape of the relation was generally linear—shoring up Hubble's expansion hypothesis—Sandage wondered whether the data points at the end of the plot veered off in a slight curvature. These data were the best available at the time, yet they were not conclusive enough to confirm any suspicions of deceleration in the expansion.

Finally, in 1956, Humason, Mayall, and Sandage published their monumental catalog of redshifts and magnitudes for 800 galaxies including Sandage's analysis. For the first time, a cache of redshifts and magnitudes substantial enough for cosmological studies appeared in a single publication. More than six decades later, this catalog is still known to astronomers as simply "HMS." From the new data, the Hubble constant H_o was recalculated and shrank to 180 km/s/Mpc, a factor of two smaller than previously.[7] And yet again, the apparent size of the universe doubled.

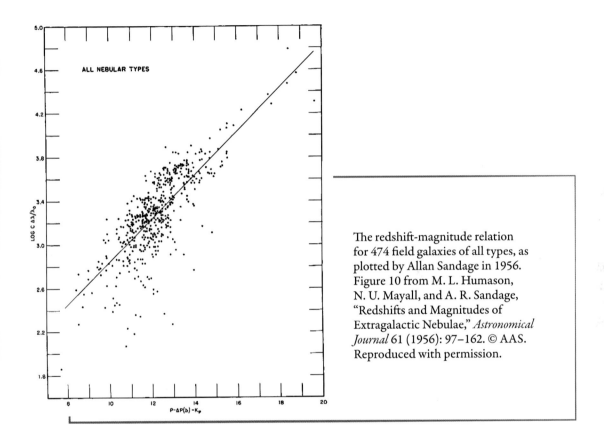

The redshift-magnitude relation for 474 field galaxies of all types, as plotted by Allan Sandage in 1956. Figure 10 from M. L. Humason, N. U. Mayall, and A. R. Sandage, "Redshifts and Magnitudes of Extragalactic Nebulae," *Astronomical Journal* 61 (1956): 97–162. © AAS. Reproduced with permission.

Eager for further success, astronomers began routinely assuming that a galaxy's brightest stars held the key to its distance. Theorists, though, expressed doubts about whether observers could reliably distinguish between isolated stars and "undefined mixtures of emission patches, star clusters, and stars."[8] Their doubts were validated by Sandage, who in 1958 laid out current problems in the extragalactic distance scale. He pointed out that some of the claimed "brightest stars" in NGC 4321, a key galaxy in the HMS analysis, were instead compact associations of bright young stars embedded in regions of ionized hydrogen. These knots were much more luminous than any individual star.[9] When all was calibrated once again, the apparent distances to extragalactic objects doubled for the third time. Sandage announced that the Hubble constant had plummeted to 75 km/s/Mpc (close to its present-day value of around 73 km/s/Mpc). The size of the observable universe had by now octupled. Would the dizzying doubling ever end?

A subject often bandied about by astronomers during the 1960s centered on the expansion of the universe: Will the universe expand forever, or will it eventually begin to collapse onto itself? This question, posed already by Edwin Hubble in 1953,[10] fired the imagination of his protégé Sandage, who solidified it over the next decade into a detailed master plan for the 200-inch telescope. His influential 1961 paper "The Ability of the 200-Inch Telescope to Discriminate between Selected World Models"[11] was the equivalent of a modern blueprint or "science case" for cosmology. It set out the theory and methods for interpreting the Hubble diagram and determining the curvature of space, which is linked to the history and future of the universe's expansion. The possibility of deciphering the origin and fate of our universe inspired many searches for standard candles that could enable the 200-inch telescope to probe deeper and deeper into space.

> "It was clear that the steps had to be much deeper than Hubble's reconnaissance—they had to be definitive. And that's what took all the time. The devil was really in the details. It looked like boring work, but it was so satisfying to get a result that one could rely on."
>
> (A. Sandage, personal interviews with the author, 2007–2010)

A great amount of telescope time was invested in determining the Hubble constant, and by the early 1960s astronomers had a better feel for the vastness of the universe. Yet controversy arose over how smooth or noisy is the cosmic expansion—known as the "Hubble flow"—as traced by galaxies. Sandage invited Gustav Tammann, a Swiss astronomer from Basel, to Santa Barbara Street to join him in what would turn out to be a decades-long collaboration to nail down the cosmological parameters. One of their first goals was to establish a Cepheid-based stepping-stone outside the Local Group of galaxies. The open arms and low surface brightness of the spiral galaxy NGC 2403 beckoned as a relatively serene backdrop for the demanding task of locating faint Cepheid variables. It did not disappoint. When the 200-inch data were analyzed, NGC 2403 turned out to be nearly five times further away from us than the Andromeda galaxy![12] Extending the Cepheid-based distance scale so far beyond our Local Group of galaxies was considered a heroic achievement at the time. Yet as Sandage and Tammann continued their campaign out to Messier 101 (known as the Pinwheel galaxy), they discovered that it was about *ten* times further away from us than Andromeda.

Measuring distances to galaxies one by one is taxing work, yet Sandage left no stone unturned in his quest to anchor the size and age of the universe. Sometimes, desperate to discover new yardsticks, he explored areas that turned out to be unfruitful. During the course of a long series of papers published with Tammann from the 1960s to the 1990s, Sandage proposed several values of the Hubble constant H_o, with errors of about 10%, all centered about H_o = 55 km/s/megaparsec. Although these values were too low by present-day standards,[13] the papers helped bring to the forefront a type of standard candle far more luminous than Cepheids.

Because the distribution of galaxies in space is so essential to our understanding of the architecture of the universe, any new potential tool for measuring distances tends to make a big wave in astronomy. Observers were cheered by Rudolph Minkowski's 1961 observation of a distant radio galaxy with a staggering redshift of 0.46. Two years later, Maarten Schmidt surpassed this record with high-redshift quasars, suddenly lurching toward the edge of the universe by a factor of 5 further than Minkowski's find.[14] Highly luminous quasars increased the hope that a reliable standard candle, independent of redshift, could be found. However, it became clear quickly that neither radio galaxies nor quasars were suitable standard candles to discriminate between cosmological models. Because they are visually indistinguishable from blue stars, quasars are difficult to identify on photographs. As they are also the extremes of a galaxy population, there are no correlations that define how intrinsically luminous they should be. So discovering the fate of the universe appeared to rest for a while on tracking down distant clusters of galaxies as possible standard candles.

The Universe Throws Down the Gauntlet

With modern detectors, rich clusters of galaxies have become visible and recognizable out to the observable horizon of the universe. However, locating them during the 1950s through 1970s was nontrivial and required close coordination between two telescopes. The Hale telescope viewed the universe through a tiny keyhole, only 0.05 square degrees (smaller than the apparent area of the moon), so searching for clusters was a crapshoot. The 48-inch Schmidt camera, with a 36-square-degree field of view, was more suited to sweeping across the sky, but clusters appeared as tiny smudges. So astronomers used the Schmidt to record large swaths of sky, then used the Hale to follow up likely suspects with photographs that reached three magnitudes fainter.

Palomar Observatory had already earned a reputation for producing high-quality cluster surveys and catalogs. One of these was Caltech graduate student George Abell's catalog of rich clusters of galaxies published in 1958,[15] a compilation of 2,400 of the richest clusters discovered on the Sky Survey plates. Another was Fritz Zwicky et al.'s *Catalogue of Galaxies and of Clusters of Galaxies.* His epic compilation of data on 40,000 individual galaxies and 10,000 clusters of galaxies, also found on Sky Survey plates, was published between 1961 and 1968 in six volumes.[16] Sandage had convinced himself that he could derive cosmological parameters from observations of the clusters' brightest member galaxies. He claimed that these ellipticals were standard candles. However, the clusters in the existing catalogs were not sufficiently remote to extend the Hubble diagram to where effects of space curvature would show up. A deeper survey was therefore undertaken by Sandage and colleagues with the 48-inch Schmidt camera and especially sensitive photographic emulsions.[17]

As data on remote galaxy clusters accumulated, warning signs began to crop up that some properties of the objects under study were evolving over cosmic time. These evolutionary effects mired attempts to turn the brightest elliptical galaxies into standard candles. In 1973 Sandage and Eduardo Hardy, a Chilean graduate student at the Hale Observatories—now Carnegie Observatories—wrote that the luminosity of the brightest galaxy in a cluster seemed to be curiously linked to both the magnitude of the second-brightest galaxy and the richness of the cluster. They named this the "Bautz-Morgan" effect after the two researchers who had classified the galaxy clusters. Not only that, but the correlation was counterintuitive: the brighter the dominant galaxy, the fainter the second- and third-ranked cluster members were. In other words, it seemed that the highest luminosity came at the expense of the fainter members. The two astronomers couched this in their paper as "The rich are rich at the expense of the poor, progressively."[18]

From his past work on stellar evolution and the color-magnitude diagrams of globular clusters, Sandage thought he understood very well how the light from aging stellar populations fades and reddens over cosmic time. Using a plot of the brightest ellipticals in 84 clusters, he assured the reader that "all the bells and whistles of the necessary corrections (K-term, galactic absorption, and population luminosity function effect) were put into the photometry."[19] Although Sandage and Hardy did not understand the basis, they nevertheless devised—and applied—empirical corrections for the Bautz-Morgan effect. What they ignored was the unwanted finding that evolutionary effects were present—and were strong enough to mask the sought-after cosmological parameters. They simply couldn't correct for the amount of luminous matter that an old galaxy can gobble up as it cannibalizes its neighbors.

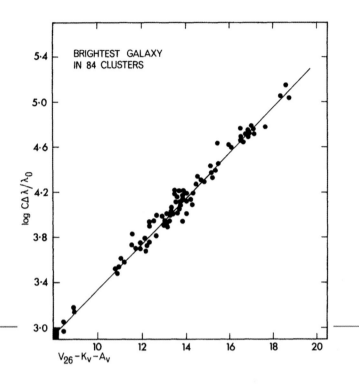

Allan Sandage hoped to discover the fate of the universe by observing the relationship between magnitudes and radial velocities of remote elliptical galaxies. Such a Hubble diagram, shown above for the brightest elliptical galaxies in clusters, was based on Palomar surveys conducted through 1971. The tiny box barely visible in the lower left corner represents the interval over which Edwin Hubble had established the redshift-distance relation in 1929. The galaxy magnitudes, plotted along the horizontal axis, are corrected for redshift and foreground dust extinction. The velocities, plotted along the vertical axis, are expressed as logarithms. Figure 4 from A. Sandage, "The Redshift-Distance Relation. II. The Hubble Diagram and Its Scatter for First-Ranked Cluster Galaxies: A Formal Value for q_0," *Astrophysical Journal* 178 (1972): 1–24. © AAS. Reproduced with permission.

Test after test of a multitude of observables later, the photometric errors remained large. In a 1999 review of his decades-long effort to nail down the geometry of the universe, Sandage characterized the final derived curvature parameter as "hardly a useful result."[20] Sadly, his effort to comprehend the universe by traditional methods had been tripped up by the brightest cluster galaxies slowly revealing themselves as unfathomably complex and unpredictable.

Chapter 2

The Whistleblower

Edwin Hubble had continued to use galaxies as simple geometrical markers to probe cosmological distances, even after Walter Baade warned his colleagues that they would not understand cosmology unless they understood the galaxies that populate space. Baade's warning took at least a decade to sink in. Afterward, the 200-inch telescope became a catalyst for the merging of theoretical and observational evidence that had been accruing. Catch-22 was about to spin a few heads around.

The first salvo came from Beatrice Tinsley, who burst onto the Pasadena scene as a postdoctoral fellow hired by Caltech professors Jim Gunn and J. Beverly Oke in 1972. As a theoretician who paid exquisite attention to detail—not being satisfied until she understood every jot and tittle—her job was to help sort out the effects of galaxy evolution on some new cluster data. Gunn, trusting Sandage's claim that all one needed to advance cosmology were the magnitudes and redshifts of the brightest elliptical galaxies in remote clusters, had joined forces with graduate student John Hoessel. In a venturesome campaign parallel to Sandage's—and expending a large amount of 200-inch telescope time—they photographed the sky at random, hoping to catch faint distant clusters in the field of view. Eventually covering more than 11 square degrees of sky, they amassed a collection of 76 distant clusters.[21]

Unorthodox and brilliant, Tinsley was known for her rigorous computer models that followed the evolution of galaxies as their stellar populations age. Her models included the photometric properties, rates of star formation, gas contents, and chemical evolution of various types of galaxies. As input for observational tests of cosmological parameters, she modeled how the apparent magnitudes and colors of galaxies change as they age.[22] However, as her work progressed, Tinsley realized that the "passive" evolution of galaxies—which is the fading of luminosity and the reddening of color that occur as generations of stars age and evolve—might not be the whole story.

During this time, papers began to appear in the literature pointing out that giant galaxies at the centers of clusters are hardly isolated systems. Rather, they interact gravitationally with other galaxies in the cluster in an ongoing process. This prompted Tinsley to ask: what happens if central galaxies grow to their monster size and brightness by swallowing some neighbors? Wouldn't that explain the Bautz-Morgan effect? Perhaps some galaxies in clusters engage in wild and sometimes violent interactions with each other. Then they brighten through accretion of matter in unpredictable ways that go far beyond an understanding based on stellar evolution. So galaxies grow dimmer with time through evolution, while they grow brighter with time through cannibalism. Such processes are not understood well enough to make precise corrections. Under these conditions, it was clearly impossible to derive any cosmological parameters from brightest cluster galaxies.

Faced with the news that cluster work was "not as advertised," Gunn's first reaction was dismay that ten years had "gone absolutely down the tubes."[23] He then realized that the powerful instruments that he and others had developed for the 200-inch—and the bounty of data collected on clusters—would be valuable for other research.

In the heart of the Virgo Cluster, an agglomeration of more than 1,000 gravitationally bound galaxies, lies a giant elliptical named Messier 87, the second-brightest member of the cluster. Such monster galaxies are among the most luminous objects in the universe. Yet they cannot serve as standard candles because the consequences of their cannibalistic behavior are difficult to predict. Credit: Sloan Digital Sky Survey.

As the astronomical community slowly accepted that galaxy clusters were a dead end as tools for measuring the universe, the focus of cosmological research advanced from a fixation with galaxies as potential standard candles to probing the evolution of the galaxies themselves. Thus, what had first appeared to be a fatal glitch instead laid the foundation for the new field of galaxy evolution. Simultaneously, Palomar gave birth to an unexpected measuring tool for cosmology, one that had been sitting in its womb for nearly half a century.

Chapter 2

"Stupendous Eruptive Phenomena"

"New stars" that appear suddenly in the sky and then fade into oblivion within a few weeks have been noted since antiquity and called "novae." However, by the early 1930s, as Fritz Zwicky and Walter Baade were investigating some apparent novae in nearby galaxies, Edwin Hubble increased the estimate of distances to galaxies from thousands to millions of light-years. As a consequence, those novae—intrinsically a thousand times brighter than previously thought—were recalibrated as "extraordinarily bright" novae and renamed "supernovae." With extraordinary insight, Zwicky and Baade hypothesized that supernovae generate cosmic rays, and that they signal the transformation of an ordinary star into a collapsed, enormously dense and small neutron star.[24]

In order to test their hypothesis, Zwicky began a systematic search for supernovae. He mounted a Wollensak camera with a 3½-inch-diameter lens on a 12-inch reflector atop the roof of the Robinson Astrophysical Laboratory on the Caltech campus. However, the initial data were so sparse that little more could be derived other than the great rarity of supernovae. Fortunately, his search caught the interest of the Observatory Council charged with building the 200-inch telescope, and they authorized the construction of the 18-inch wide-field Schmidt camera. As soon as it was commissioned in 1936, Zwicky and his assistant began using it every night, while Baade made photometric follow-up measurements of the waning light of the discovered supernovae at Mount Wilson Observatory.

By 1938, Baade, having compiled sufficient photometric data for light curves of 18 supernovae, wrote that these extraordinarily luminous objects had similarly shaped light curves.[25] Three years later, Mount Wilson spectroscopist Rudolph Minkowski announced that, of the 14 objects for which he had obtained spectra, five had light curves that differed considerably from those of the supernovae studied by Zwicky and Baade. Also, strong Balmer emission lines of hydrogen appeared in their spectra soon after maximum light. Minkowski provisionally designated these as "Type II" supernovae. The remaining nine objects formed an "extremely homogeneous group," with spectra showing broad emission bands, no hydrogen lines, and only two "forbidden" (rarely observed) bands of oxygen. He provisionally named these objects, which resembled the ones analyzed by Baade, "Type I" supernovae, and recorded one light curve from 7 days before maximum to 339 days after. The evolution of the light curves and spectra of Type I supernovae were so similar that Minkowski could tell how much time had elapsed since maximum from their luminosities and spectra alone.

By the time Minkowski wrote a review paper in 1964,[26] 140 supernovae had been discovered in searches directed by Zwicky. Although Minkowski further expanded the base of empirical light curves for Type I supernovae, he still found no satisfactory explanation for the supernovae's spectra. He simply described them as consisting of broad features that could represent emission from a hot gas with a source of energy, absorption of light from a hot source passing through cooler material, or a mixture of both.[27]

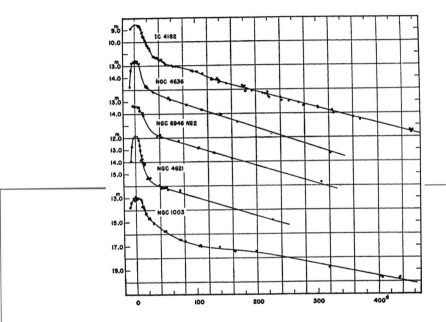

Rudolph Minkowski's photographic light curves for five Type I supernovae found by Zwicky. Before Minkowski's work, few spectra of supernovae had been photographed, and the only information on spectra came from visual observations. For more than three decades, Minkowski's detailed recording of the spectrum and light curve of SN 1937C would provide the main observational basis for attempts to empirically analyze supernova phenomena. Figure 1 from R. Minkowski, "Spectra of Supernovae," *Publications of the Astronomical Society of the Pacific* 53 (1941): 224.

Despite the irregular sampling of their light curves, the large uncertainties in the photographic photometry, and the probable light contamination from host galaxies, evidence slowly accumulated that supernovae of Type I might one day become robust cosmological tools.

CHAPTER 2

CINDERELLA IN THE SUBBASEMENT

At their maximum, supernovae can be 4 million times brighter—and visible two thousand times further away—than the Cepheids Hubble used to measure the distance to M31. However, the potential of supernovae as cosmological standard candles was not fully unleashed until a skinny, quiet, and very dedicated young man, with an undergraduate degree in astronomy from the University of Southern California, published an outright supernova Hubble diagram. Charles Kowal was hired in 1963 as Palomar's "supernova hunter" to continue the project after Zwicky's retirement. These were the early days when searches were made by visual inspection and comparison of pairs of photographic plates taken during the dark of the moon. It was tedious labor, yet sometimes exciting, and Kowal did this year after year for a quarter of a century.

Earlier attempts by Baade and Minkowski to measure the peak magnitudes of supernovae had been thwarted by the scarcity of observations made before maximum light. In his office in a subbasement of Robinson Hall, Kowal analyzed "reasonably reliable" estimates for the peak photographic magnitudes of 40 supernovae of various types that had been discovered between 1895 and 1967. After making a few of the usual corrections for Milky Way foreground absorption and solar motion, he estimated the distances and peak absolute magnitudes for 19 Type I supernovae for which light curves were available. Noticing that their peak absolute magnitudes, and hence intrinsic peak luminosities, were remarkably uniform, he then plotted their peak apparent magnitudes against their redshifts in the classic diagram that Edwin Hubble had originally devised for elliptical galaxies.

> During an interview with *Time* magazine, Charlie Kowal made an offhand remark about the politics of his below-ground office in Robinson Hall at Caltech. It was then quoted in the October 27, 1975 issue: "This building is like an ocean liner. All the professors are up there on the promenade deck promenading. This is the engine room where all the work gets done." Fritz Zwicky's office was also in the subbasement, and the two supernova hunters were good friends.
>
> (C. Kowal, personal interview with the author, 2010)

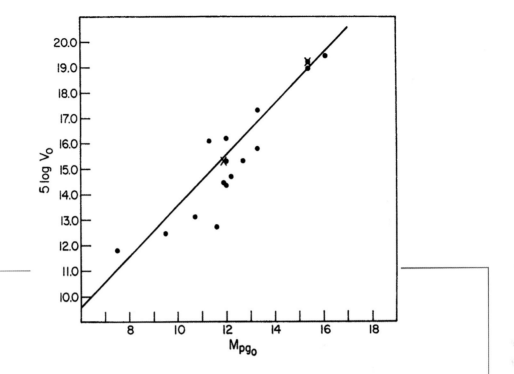

The first Hubble diagram for supernovae, plotted by Palomar supernova hunter Charlie Kowal in 1968. The dots represent individual supernovae of Type I observed in galaxies out to recession speeds of 7,000 km/s. The two crosses represent the average velocity and supernova peak magnitude in the Virgo and Coma clusters of galaxies. Kowal fitted the crosses with a 45-degree straight line (representing a linearly expanding universe) and estimated an accuracy of ± 0.3 magnitudes for this mean relation. Figure 1 from C. T. Kowal, "Absolute Magnitudes of Supernovae," *Astronomical Journal* 73 (1968): 1021–1024. © AAS. Reproduced with permission.

The scatter of individual galaxies around the mean relation, although admittedly large—with a dispersion of about 0.6 magnitudes (or 30% in distance)—is due to measuring errors and variations in the internal absorption of light (termed extinction) within the host galaxies. In spite of all that, in a 1968 paper Kowal predicted that the scatter could one day be narrowed to the point that Type I supernovae would become reliable cosmological standard candles.[28] He even guessed that, by adding very distant Type I supernovae to the Hubble diagram, any subtle curvature in the mean relation might show up. If the magnitude-redshift relation turns upward at large distances, it means that the expansion of the universe is accelerating. If it turns downward, the expansion is decelerating. In either case, Kowal forecast that the mean peak brightness of supernovae in galaxy clusters would be eventually measured to an accuracy of 0.1 or 0.2 magnitudes—corresponding to 5% to 10% in distance. In contrast to the abandoned idea of measuring distances based on brightest galaxies in clusters, supernovae appeared relatively free of complicated evolutionary effects while still being visible at large distances. So, despite using noisy photographic data and primitive techniques, Kowal had a keen nose for the future.

Enter Robert Kirshner, a new graduate student at Caltech in 1970. When J. Beverly Oke, his thesis advisor, asked him what he wanted to work on, Kirshner mentioned that he had enjoyed working on the Crab Nebula. Oke then opened his desk drawer and pulled out a stack of supernova spectra that nobody had ever measured—or even looked at. It seems that the faculty kept taking spectra of supernovae with the 200-inch telescope because it was a tradition and an adventure, and they hoped that, someday, something would be done with them. Here was Oke's chance to pass on several years' worth of accumulated spectra. He handed Kirshner the stack of glass plates wrapped in yellow envelopes and Kirshner said "thank you," but he had no idea what to do with them or how to do it.[29] As was the tradition, other students came to his rescue.

During this time, Kowal was running a program on the 18-inch Schmidt to look for supernovae in nearby spiral galaxies. On May 13, 1972 he caught a bright 8th magnitude supernova just after it reached maximum brightness, in the spiral galaxy NGC 5253. Named SN 1972E, it was located near the periphery of the galaxy, where the background interference was low. Kirshner raced to Palomar to track the supernova's rapidly changing spectrum, hoping to bring order to the seemingly haphazard behavior of spectral features in supernovae. To maximize the number of spectra he could take, he alternated between observing at the 200-inch telescope and the 60-inch reflector that had been installed on the mountain in 1970. At the 200-inch, Kirshner used Oke's newly invented spectrophotometer, called the "multichannel" because it simultaneously measured the light from an object at 32 different wavelengths instead of just one. It was among the first instruments to record spectra digitally and provided a spectral atlas of the supernova from 3,300 to 10,000 angstroms.

Prior to the observations of SN 1972E,[30] the only comprehensive observations of a supernova were the prismatic spectra of SN 1937C, a Type I supernova, taken by Minkowski[31] thirty-five years earlier. The spectra of the two supernovae were nearly identical and evolved at the same rate for at least 225 days after maximum light. Thus, both were Type I supernovae. The high resolution of the multichannel spectra allowed Kirshner and colleagues to identify emission lines of forbidden iron, ionized calcium, sodium, magnesium, and nickel that made up part of the nebular spectrum. They also discovered an envelope of material expanding at 20,000 km/s in the outer regions of the supernova. However, a deep understanding of the physics of phenomena such as the continuous source of energy that fueled the expanding envelope still eluded them. It would take the expertise of many dedicated researchers more than two decades to understand the blending of emission and absorption lines and the "dropping out" of some spectral features over time.

Certainly, the class of Type I supernovae would need to undergo several refinements before supernovae could be elevated to the status of "standard candle." For example, Sandage and Tammann assured their colleagues that the Type I supernovae they had observed showed no systematic deviations from template light curves, and that their peak luminosities were nearly identical. Yet Russian astronomer Yuri Pskovskii reached the opposite conclusion. After collecting published data on Type I supernovae, many from Palomar, he arranged their light curves according to how fast the light faded. Bucking the claims made by other researchers, in 1977 Pskovskii asserted that there were significant intrinsic differences among the peak luminosities of Type I supernovae. Their light curves also declined at different rates, he wrote. Furthermore, the rate of decline correlated with several parameters, the most interesting being the peak luminosity: brighter Type I supernovae faded slowly, and fainter supernovae faded quickly.[32] Although that correlation would turn out to be a highly significant finding for the cosmological program, it took more than a decade for Pskovskii's results to be confirmed. To be fair, the delay was partly due to the uncertain quality of the motley collection of photographic data and incomplete light curves he had culled from the literature.

The debate over the homogeneity of the Type I class of supernovae continued, with some astronomers claiming intrinsic differences in their optical spectra and light curves, while others dismissed any differences as observational error. So far, the dispersion in peak magnitudes of Type I supernovae at visible wavelengths was claimed to be a few tenths of a magnitude. But was the dispersion intrinsic to the supernovae? Or was it due to differences in interstellar extinction, the dimming of starlight due to absorption and scattering within the host galaxies? One thing astronomer Jay Elias knew very well was that any such extinction would be several times smaller at infrared wavelengths than at visible wavelengths, since galactic dust absorbs much less infrared radiation than visible. As the sensitivity of infrared detectors had vastly improved since Kirshner's and Oke's pioneering observations of SN 1972E, Elias and three colleagues were able to measure the light curves of 11 Type I supernovae with the 200-inch Hale telescope. They worked in the infrared, at 1.1 microns (the J band), 1.6 microns (the H band), and 2.2 microns (the K band).

What they found, and described in their 1985 paper,[33] was completely unforeseen. Some of the light curves they constructed in the infrared were fundamentally different from the characteristic shapes of the light curves Minkowski, Kirshner, Pskovskii, and others had constructed at visible wavelengths. In the infrared, only a minor number of Type I supernovae exhibited a normal, single-maximum light curve. Most of the Type I supernovae went through a *second* brightening, and it was sometimes brighter than the first. In every case, the supernovae reached maximum light in the H and K bands a few days earlier than in the J band. There appeared to be two well-defined subclasses of Type I supernovae, which could be distinguished from the shapes of their infrared light curves. Elias designated the supernovae with two peaks in their light curves as Type Ia, and the ones with a single peak as Type Ib. Whereas the earlier distinction between Type I and Type II supernovae made by Minkowski in 1964 was on the basis of spectroscopy, this new distinction between Type Ia and Type Ib was made by photometric classification. It was based on the shape of the light curve.

Elias and colleagues plotted a Hubble diagram restricted to the new class of Type Ia supernovae, but they used the peak infrared magnitudes rather than the peak optical magnitudes. Although they didn't have many points for their plot, the points fit on a line well enough to suggest that Type Ia supernovae are nearly perfect standard candles in the infrared. The dispersion in peak magnitudes appeared to be less than 0.2 magnitude and possibly, the team wrote, less than 0.1 magnitude!

Optical astronomers had yet to catch up. With CCD (charge-coupled device) detectors steadily improving, astronomer Mark Phillips of the Cerro Tololo Inter-American Observatory in Chile was intrigued enough to reexamine Pskovskii's early claims. Using precision photometry of supernovae from the past decade, he found that the peak optical magnitudes of nine Type Ia supernovae varied significantly. He also confirmed the existence of a strong correlation between peak magnitudes and decline rates. By codifying this correlation, he "standardized" Type Ia supernovae to serve as reliable standard candles out to cosmological distances. It is impressive that Kowal, with the rather primitive photographic data of his time, had accurately predicted how the field would progress. After some further corrections, Type Ia supernovae turn out to be among the most precise cosmological distance indicators known, and much effort is being devoted to finding many more of them. Currently, their distances can be measured individually to an accuracy of better than 10% out to around 10 billion light-years.[34]

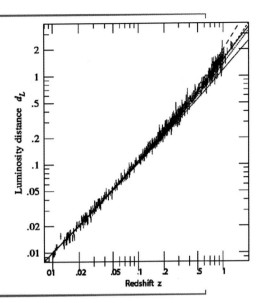

The Hubble diagram, long hoped to one day represent the history of deceleration of the universe, sprang a great surprise on astronomers. For reasons unknown and unpredicted, the universe switched its expansion rate from decelerating to accelerating around 5 billion years ago, which remains a profound mystery. In the current Hubble diagram, the uptick representing acceleration of the expansion begins at around redshift $z = 0.5$. After figure 8 from M. Betoule et al., "Improved Cosmological Constraints from a Joint Analysis of the SDSS-II and SNLS Supernova Samples," *Astronomy and Astrophysics* 568 (2014).

In retrospect, astronomers stumbled in their bold attempt to measure the curvature of space and the mass of the universe. Their early goal of determining the nature of the universe was unachievable by the means originally envisioned, and their methods were surpassed by others then not even on the horizon. Yet observations with all four Palomar telescopes—the 18- and 48-inch Schmidts, and the 60- and 200-inch reflectors—laid the groundwork for a precision measuring tool that is apparently immune to evolutionary changes over cosmic time. Eventually, six decades after Fritz Zwicky began scanning the sky for supernovae, and five decades after Edwin Hubble ventured that the law of redshifts might show whether the expansion of the universe has been slowing down or speeding up, the contrarian universe yielded its answer. Two fiercely competing teams of astronomers spent years taking observations with ground-based telescopes, including the two Keck 10-meter telescopes on Maunakea and the Blanco 4-meter telescope at Cerro Tololo, and with the Hubble Space Telescope. Although the teams were independent, their goals were parallel: to lock down the parameters that describe the expected gradual slowing of the universe's rate of expansion due to the mutual gravitational force between objects in the universe. After much calibration, analysis, and sharp debates, the two teams separately published their results in 1998[35] and 1999,[36] announcing one of the most baffling discoveries of the 20th century. Sometime in the past five or so billion years, the expansion rate of the universe switched from deceleration to *acceleration*. For this discovery, the two teams shared the Nobel Prize in Physics in 2011.

Hanging in Allan Sandage's office, the diagram that made him famous. Photo by the author.

· 3 ·
Unraveling the Mysteries of Stellar Evolution

Chapter 3

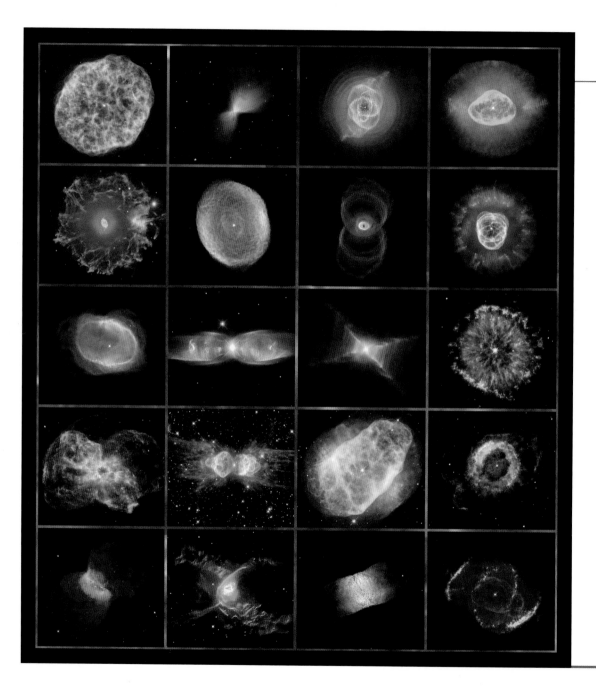

Our Milky Way galaxy is littered with stellar relics called planetary nebulae. They are the brilliantly artistic sepulchers formed by dying stars during the final evolutionary stages of their lives. Such a nebula forms from the abundance of gas shed by winds, flares, and storms that are active in the atmosphere of a red giant star as it evolves into a white dwarf. At first, the gas glows brightly from the ionizing radiation of the hot stellar core, often visible as a white dot in the center of the nebula. The myriad shapes and colors of the nebulae depend on whether the stars are binary, rotating, the degree of chemical enrichment, the strength of magnetic fields, and the presence or absence of a planetary system. However, beauty is fleeting. The planetary nebula slowly ghosts away over some thousands of years as it dissipates into the Galaxy. The late stages of stellar evolution play a role in dispersing the products of nucleosynthesis, gradually enriching the chemical content of the universe, from which new stars, planets—and life—are continuously formed.

Credits, top row: ESA/Hubble and NASA, acknowledgment: Marc Canale • NASA, ESA and The Hubble Heritage Team (STScI/AURA) • ESA, NASA, HEIC, and The Hubble Heritage Team (STScI/AURA) • Bruce Balick (University of Washington), Jason Alexander (University of Washington), Arsen Hajian (U.S. Naval Observatory), Yervant Terzian (Cornell University), Mario Perinotto (University of Florence, Italy), Patrizio Patriarchi (Arcetri Observatory, Italy), and NASA/ESA.

Second row: Nordic Optical Telescope and Romano Corradi (Isaac Newton Group of Telescopes, Spain) • NASA/ESA and The Hubble Heritage Team (STScI/AURA) • Raghvendra Sahai and John Trauger (JPL), the WFPC2 science team, and NASA/ESA • NASA, ESA, Andrew Fruchter (STScI), and the ERO team (STScI + ST-ECF).

Third row: The Hubble Heritage Team (STScI/AURA/NASA/ESA) • Bruce Balick (University of Washington), Vincent Icke (Leiden University, the Netherlands), Garrelt Mellema (Stockholm University), and NASA/ESA • NASA/ESA, Hans Van Winckel (Catholic University of Leuven, Belgium), and Martin Cohen (University of California) • NASA/ESA and The Hubble Heritage Team (STScI/AURA).

Fourth row: NASA, ESA, and K. Noll (STScI) • NASA, ESA, and the Hubble Heritage Team (STScI/AURA) • ESA/Hubble and NASA, acknowledgment: Matej Novak • NASA/ESA and The Hubble Heritage Team (STScI/AURA).

Bottom row: NASA/ESA and The Hubble Heritage Team (AURA/STScI) • ESA and Garrelt Mellema (Leiden University, the Netherlands) • NASA/ESA and The Hubble Heritage Team STScI/AURA • J. P. Harrington and K. J. Borkowski (University of Maryland) and NASA/ESA.

Image composition by B. Schweizer.

CHAPTER 3

Often, the search for patterns and correlations between various properties of objects is a first step toward understanding processes. As an example, in the late 19th century, when the chemical elements were ordered by their atomic number, mass, electron configuration, and chemical properties, they fell into groups with similar attributes, such as degree of reactivity, metals versus nonmetals, gases versus solids, and so on. The resulting periodic table of elements provided a tool for figuring out the behavior of chemical reactions. Similarly, in the early 20th century, when measured colors of stars were plotted against their magnitudes, most stars fell into a diagonal swoosh that stretched from the region of cool faint red stars to the region of hot bright blue stars. Now referred to as the "main sequence," this swoosh represents a relatively calm and long-lasting equilibrium stage in stars' lives, during which energy is generated in their cores through the nuclear fusion of hydrogen into helium. Of special interest are color-magnitude diagrams for globular star clusters, which are tightly bound aggregations of many thousands to millions of stars in a galaxy's halo. These diagrams are one of astronomers' prime tools for understanding how stars evolve, for rectifying the cosmic distance scale, and for understanding stellar nucleosynthesis.

THE BREAKTHROUGH

In the 1940s, Walter Baade had eagerly awaited his first observing run on the 200-inch telescope. Besides scrutinizing Edwin Hubble's questionable distance to the Andromeda galaxy, his master plan included learning more about the two stellar populations he had discovered there. The first step was to search for main-sequence stars in a few of the nearest globular clusters in our Milky Way galaxy. The beauty of globular clusters is that they act as petri dishes for studying stellar evolution. Pre-Palomar attempts to observe main-sequence stars in them were pushed to the limits of existing photometric magnitude scales and photographic emulsions. Yet main-sequence stars had not been detected in even the nearest globular clusters. The much brighter red giants had been observed, but the connection between the two classes of stars was not yet understood.

Baade focused on the globular cluster Messier 3 (M3)—a spherical swarm of half a million colorful stars jammed into a few arcminutes in the northern sky. He took deep photographs in blue (B) and yellow ("visual" V) light and handed the plates to graduate student Allan Sandage for his PhD thesis project. Sandage's ultimate goal was to determine the absolute magnitudes of M3's RR Lyrae variable stars. These were the primary so-called "standard candles" available in old stellar populations for measuring large distances in the Milky Way and to neighboring galaxies. For this task, he needed to discover and map M3's main sequence, and the region where the RR Lyrae variables reside, in the color-magnitude diagram. Once he had found the main sequence, Sandage compared it to the location of the main sequence of stars in the solar neighborhood, whose distances were known from stellar parallaxes. Finally, the distance to M3 was derived from the offset between the two main sequences. Neither Baade nor Sandage had an inkling that their work would soon have much wider consequences.

Sandage began this tedious project by marking the positions of 1,100 stars on each of Baade's many blue and visual plates, avoiding areas with heavily blended stellar images. He worked relentlessly for a year in the windowless subbasement of the observatory headquarters on Santa Barbara Street in Pasadena. Since the existing sequences of "standard star" magnitudes did not extend faint enough to reach the limit of the 200-inch telescope, Sandage bridged the brightness gap with a so-called "fly spanker." This was a small glass photographic plate on which a star of known magnitude was exposed for a series of time-intervals corresponding to fixed steps in magnitude. Since the plate was shifted incrementally between exposures, the result was a series of dots of decreasing size and density that approximated the appearance of stars of various magnitudes out to the faintest stars to be observed. With the aid of the existing sequences and the fly spanker, Sandage measured 5 plates in each color for all stars in his sample. He then repeated the measurements three times to estimate probable errors in the final magnitude and color of each star. As he plotted his measurements by hand on a color-magnitude diagram, a diagonal pattern slowly emerged.

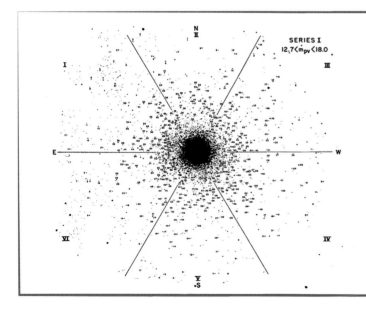

One of Sandage's identification charts for the globular cluster Messier 3, made from an enlarged print of a 200-inch prime-focus plate and with all stars measured marked with numbers. Figure 6 from A. R. Sandage, "The Color-Magnitude Diagram for the Globular Cluster M 3," *Astronomical Journal* 58 (1953): 61. © AAS. Reproduced with permission.

Chapter 3

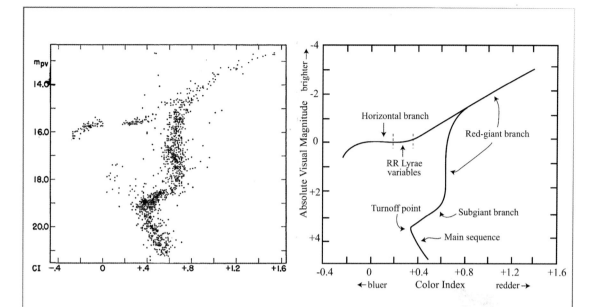

Left, Sandage's 1953 color-magnitude diagram for stars in the globular cluster Messier 3. Each point represents one of 1,100 stars he selected for measuring from the cluster's estimated half-million stars. Each star's color index is plotted along the horizontal axis, while its apparent visual magnitude is plotted along the vertical. Since stars in a globular cluster are likely to have formed from a single gas cloud, they share similar chemical compositions and ages. Thus, the location of a star in the color-magnitude diagram reveals its evolutionary state, which depends only on its mass and age. Low-mass stars evolve slowly, while high-mass stars burn out relatively quickly in cosmic terms. The main sequence of Messier 3, visible as the short stub at the bottom of the plot, represents the first stage in a star's life. During this stage, which is relatively calm and long-lasting, energy is generated through the thermonuclear fusion of hydrogen into helium. Stars that have consumed their central hydrogen fuel change in structure and grow larger, brighter, and redder as they evolve toward the "giant branch" in the upper right of the diagram.

Right, a sketch of M3 with the various evolutionary stages labeled. Stars more massive than those at the main-sequence turnoff are evolving up the subgiant branch to the red giant branch and then, after complex evolutionary processes and mass loss through stellar winds, are populating the horizontal branch. There, during their evolution from the right to the left, they spend some time pulsating as RR Lyrae variables. After reaching the blue end of the horizontal branch, they become white dwarfs and fade away.

Left, figure 1 from A. R. Sandage, "The Color-Magnitude Diagram for the Globular Cluster M3," *Astronomical Journal* 58 (1953): 61. © AAS. Reproduced with permission. Right, after figure 4 from W. Baade, *Evolution of Stars and Galaxies* (Cambridge, MA: Harvard University Press, 1963).

In the fall of 1951, after plotting his last few data points, Sandage climbed the stairs to Baade's office carrying the large sheet of paper under his arm. Baade's good friend, renowned theoretical astrophysicist Martin Schwarzschild, happened to be visiting from Princeton University and looked on as Sandage held up his plot. The three men first noticed that, for his protracted efforts, Sandage had indeed located the main sequence of the cluster. It appeared as a well-defined and densely populated but rather stubby mass of points grouped near the bottom of the plot, where the faintest stars lay. Surprisingly, at brighter magnitudes it ended abruptly by turning off to the right and making a beeline for the region occupied by red giant stars. The realization that there were no stars on the main sequence brighter than this turnoff point made the two senior astronomers' jaws drop. The well-defined limb that jutted toward the giant region strongly suggested to them that as a cluster ages, its stars evolve *off* the main sequence and *toward* the region of giant stars. The complex pattern of the color-magnitude diagram suddenly made sense!

As Schwarzschild scrutinized the form of Messier 3's color-magnitude diagram, it struck him that nature herself had breached a theoretical impasse that had confounded him for years. Baade had discussed the subject of stellar interiors with him, but merely as a side issue. However, the moment Sandage held up the plot, the interests of the East Coast theoretician and the West Coast observer merged. Along with resolving the long-standing empirical dilemma of the distance scale, the plot promised to revolutionize the theory of stellar evolution.

The inferred movement of stars toward the giant region was a revelation for two reasons. First, prior to Palomar most astronomers believed that stars began their lives as red giants, then lost mass and descended in the color-magnitude diagram toward the main sequence. To the contrary, Sandage's color-magnitude diagram of Messier 3 suggested that main-sequence stars may evolve *into* red giants. Second, theoreticians of the time had not been able to solve the equations of stellar structure and energy generation required to model the evolution of a stable, hydrogen-burning star as it leaves the main sequence. Sandage's discovery of a connection between the main sequence and giant stars showed theoreticians that a solution was possible.

Sandage's globular-cluster data intrigued Scharzschild, who was hatching ideas to explain the startling morphology of the color-magnitude diagram. He convinced Baade that Sandage should work with him to develop new models, and the young researcher went to Princeton on a fellowship to study "what had been seen through the world's largest telescope."[37] Since electronic computers were not yet generally available, Sandage built stellar models by performing lengthy and complex numerical calculations on electromechanical machines by hand. His detailed computations indicated that when a main-sequence star has fused about 12% of its hydrogen mass to helium, the thermonuclear reactions in its core cease since the temperature is too low to ignite helium burning. Schwarzschild had been plagued for years by the instability of this configuration, which represented an impasse to calculating the star's further evolution. By studying the full extent and complexity of Sandage's color-magnitude diagram, Schwarzschild finally grasped the modus operandi of stars leaving the main sequence.

Schwarzschild and Sandage continued their computations. With diminishing energy generation and radiative support, the model star's core slowly contracted, liberating gravitational energy that heated a surrounding shell of hydrogen and increased the rate at which thermonuclear reactions proceeded. The infusion of heat from the hydrogen-burning shell expanded the outer layers of the star, which expanded into a red giant. After some further evolution, the model star consisted of a series of layers resembling an onion: an inert core of heavy elements surrounded by a helium-burning shell, then a helium-rich intermediate zone surrounded by another, hydrogen-burning shell, with this entire structure wrapped in a hydrogen-rich radiative envelope. The two researchers' computations suggested that as thermonuclear reactions continued, the bloating star rapidly became redder and brighter as it evolved to the upper right of the color-magnitude diagram, called the giant branch. Although there were still significant uncertainties, this model, published in 1952,[38] seemed to reproduce the main features of Sandage's color-magnitude diagram for Messier 3 reasonably well.

A windfall of Sandage's discovery of the main-sequence turnoff came in the form of a recipe for estimating the ages of globular clusters from this turnoff. As cluster stars age, the most massive ones—which burn their fuel fastest—are the first to exhaust their hydrogen supply and evolve off the main sequence. Thus, at any given time, the brightest stars found at the turnoff have burned about 12% of their hydrogen and are ready to move on to the red giant stage. The mass of these stars can be calculated from their luminosity. Then, using the known rate at which mass is converted from hydrogen to helium, Sandage and Schwarzschild calculated how long the star at the turnoff must have been burning hydrogen. This, then, yielded directly the age of the cluster.

By the mid 1950s, Sandage amassed color-magnitude diagrams for ten open clusters, in addition to the diagram for Messier 3. Compared to a globular cluster, such open clusters are relatively sparse, with typically a few dozen to at most a few thousand stars in them. Also, they tend to lie in the disk of the Milky Way, where destructive tidal forces may be strong enough to destroy them. Hence, there are few very old open clusters. Yet Sandage noticed a similarity and apparent continuity between the color-magnitude diagrams of open clusters of various ages and that of the globular cluster Messier 3. Superposing his eleven color-magnitude diagrams with their stellar main sequences matched, he demonstrated that they form an *age sequence*. This diagram became a textbook favorite and made Sandage famous.

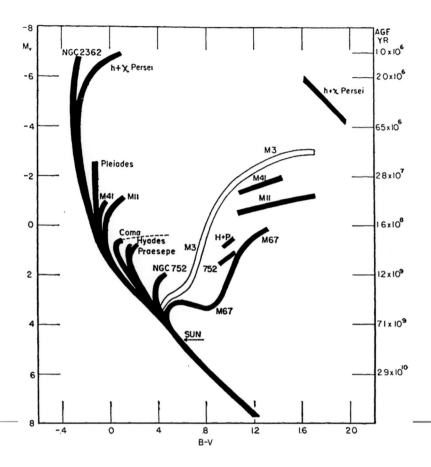

The composite color-magnitude diagram for 10 open clusters and the globular cluster Messier 3 displays the absolute visual magnitudes of cluster stars, M_v, plotted against their color indices (blue–visual, a proxy for a star's surface temperature). Brightness increases upward and temperature increases toward the left in the diagram. The position of a star along the main sequence depends only on the star's mass, and Sandage's composition demonstrates that the strongest variable is time. If a cluster is observed early in its life, the main sequence is populated by stars along its entire length. For example, NGC 2362 and h and χ Persei are so young that only their most massive stars have begun to leave the main sequence. If you catch a cluster later, many of its stars have wandered off to the giant branch, and the main sequence appears stubby (for example, in Messier 67). Figure 1 from A. Sandage, "The Systematics of Color-Magnitude Diagrams and Stellar Evolution," *Publications of the Astronomical Society of the Pacific* 68 (1956): 498. © The Astronomical Society of the Pacific. Reproduced by permission of IOP Publishing. All rights reserved.

Chapter 3

Allan Sandage, a self-described "astronomy junkie," works at his desk among his books, toys, and models. Photo by author.

"During the period of explosive discovery at Palomar that lasted from 1950 to 1965, I was psychologically in a different world … part of something fundamentally magnificent that was so much bigger than everyday life. It's the way I suppose religious people would describe some sort of a revelation they have had … some sort of epiphany. It's never happened to me in religion, but it certainly did occur for 15 years straight in my research. Everything was opening, and it was like picking flowers in so many fantastic gardens.

"Stellar evolution is really closest to my heart, and understanding the topology of the color-magnitude diagram with the movement off the main sequence was just such a time. I was living on all 10 cylinders at once. Every day, every week, every month. I remember that feeling was so … powerful. I spent 100 nights per year on 3 mountains for 30 years. So, it was completely out of this world. It was being in Xanadu and the pleasure dome of Kubla Khan. [Chuckles] … Every day, every day."

(A. Sandage, personal interviews with the author, 2007–2010.)

Sandage included the globular cluster Messier 3 in his comparison to make the point that there may be unknown parameters at play in the morphology of its color-magnitude diagram. For example, the shapes of the giant branches of M3 and M67 differ substantially, even though the two clusters are nearly the same age, as judged from their main-sequence turnoff points. Since M67 is close to the galactic disk and M3 lies far above in the halo, the chemical compositions of their birthplaces may have been very different. The fact that not only the age but also the chemical composition of a cluster influences its color-magnitude diagram added a level of complexity. By 1970, it was clear that age affects color-magnitude diagrams more than chemistry, but exactly how? Again, work done at the 200-inch telescope—this time in the near infrared—paved the way. The advantage of observing in the near infrared was that a giant star's luminosity and temperature could be measured much more accurately there, while at the same time reddening due to foreground dust was diminished. By measuring the near-infrared colors of stars, Jay Frogel, Judith Cohen, and Eric Persson showed that different types of stars—some of them extremely luminous, but prominent mainly in the infrared—populate the giant branch of 26 globular clusters.[39] Thus, in the infrared, cluster ages and chemical compositions can be studied with higher precision and less interference from dust clouds in the Milky Way.

Just how narrow is the distribution of stars along the main sequence? The transition from photographic plates to photoelectric detectors, such as CCDs, in the 1980s showed that most of the scatter around the main sequence had been observational. At present, observations from space satellites like the Hubble Space Telescope show that stars line up along the main sequence like beads on a string.

Stellar Radioactivity, Storms, Flares, and Winds

Scientific progress often occurs in parallel instead of sequentially, and scientific results rarely arise in vacuo. In the 1950s, astronomers were occupied mostly with classical astronomy—setting up magnitude scales, photographing galaxies for classification, assembling catalogs of galaxy redshifts and magnitudes, constructing color-magnitude diagrams for stars, and calibrating standard candles to refine the distance scale and the Hubble constant. The idea that supernova explosions enrich the interstellar medium with heavy elements was still relatively new, and aging stars blowing out winds laden with dust had yet to be discovered. Most astronomers believed that heavy elements formed in the big bang, and that the universe had a homogeneous chemical composition. That changed over the next decade as a seemingly unrelated clutch of observational, theoretical, and experimental insights convinced most astronomers that stars not only create chemical elements but also distribute them on a cosmic scale.

Having passed the age of compulsory retirement, world-class Mount Wilson spectroscopist Paul W. Merrill was prohibited from observing at Palomar. So Palomar director Ira Bowen took spectra of eight cool red giant stars on his behalf during the initial tests of the 200-inch telescope's high-dispersion coudé spectrograph. The spectrograms were extraordinarily complex and crowded, with many spectral lines severely blended. In fact, the atmospheres of red giants are so laden with molecules that their many exceedingly weak atomic and molecular lines had not yet been identified in the laboratory. In spite of such obstacles, Merrill accomplished the stunning feat of recognizing that some of the feeble lines were due to an unexpected radioactive element. In a 1952 paper notable for its understatement and restrained title, "Spectroscopic Observations of Stars of Class S," he announced the detection of technetium in red giant stars. This was a significant claim: if technetium was synthesized in the big bang, even its longest-lived isotope, with a half-life of around 4 million years, would have decayed long before the red giant was even born. Thus, Merrill's finding was the first direct evidence that sites of ongoing nucleosynthesis occur deep inside some types of stars, and mixing or convection then transports the products to the surface.[40] Due to the fundamental importance of this discovery and to some lingering reservations about the correctness of his tricky identification, others set to work and confirmed his claims.

Mira, a high-velocity red supergiant star in the late stages of evolving into a white dwarf, leaves a long wake of cast-off material and creates a gaseous bow-shock in front as it races through the interstellar medium.
Credit: NASA/JPL-Caltech.

There were more odd things going on in the atmospheres of supergiant stars. For example, a succession of observers at Mount Wilson Observatory had reported that absorption lines in the type M supergiant α Herculis were asymmetric, with narrow, deep troughs on the blue side of the lines. At first, they interpreted this phenomenon as due to gaseous convection currents in the stars' atmospheres—fountains of gas that rise above the star's surface only to fall back because their velocities aren't high enough to escape the star's gravity.

This interpretation was challenged by Mount Wilson astronomer Armin J. Deutsch, who in 1956 took 200-inch coudé spectra of the supergiant—and shrewdly included its fainter companion, a G giant star. As the same blue absorption troughs were visible in the spectra of both stars,[41] he concluded that the absorption lines were due to a single enormous expanding cloud that enveloped both stars. Driven by unknown forces, gas was clearly streaming out of α Herculis and into its surroundings, enveloping its binary companion hundreds of astronomical units away, with no signs of falling back into the supergiant. With this decisive result, Deutsch appeared to have stumbled upon a characteristic of all M supergiants: they pour mass into their surroundings. Red giants and supergiants become so extended—and their surface gravities so extremely low—that bits of their atmospheres can be thrown from the surface into space by storms, flares, and winds.

Theoreticians were pleased with Deutsch's discovery, since their models had suggested that massive stars, having exhausted their hydrogen fuel during the red giant phase of evolution, would likely eject material. This blowing away of the outer layers may be important in the late-stage evolution of all massive stars. Some aging stars violently explode, while others softly puff away their dusty smokestack.[42]

Fired up by Deutsch's discovery, Dieter Reimers, an astronomer at Kiel University, reanalyzed hundreds of spectrograms of giant and supergiant stars borrowed from the collections that Deutsch and others had stored in the plate vaults of the Hale Observatories. As he pored over the many spectrograms, Reimers realized that each of the previous estimates of mass loss in red giants and supergiants had referred to a different height in the stars' expanding envelope. In a flash of insight, he reasoned that the rate of mass loss should depend on parameters that define the *structure* of stellar atmospheres—gravity g, luminosity L, and radius R—and their chemical composition. In addition, he examined the potential influence of stellar rotation, magnetic fields, chromospheric activity, and pulsation. As he assembled and plotted all the available data and resolved some uncertainties in Deutsch's work, Reimers demonstrated that giant stars blow out a thousand times more material than previously thought.[43] Thus, not only exploding stars but millions of aging giant stars in the Milky Way are significant sources of chemical enrichment of the interstellar medium.

Nucleosynthesis Enshrined

One of Fritz Zwicky's passions was surveying for supernovae—stars that explode violently at the end of their life. Thus, it seemed odd that he and Walter Baade chose to include the anemic dwarf galaxy IC 4182 in their supernova search, since there was little chance that such a sparse galaxy would produce a supernova. Yet against the odds the galaxy soon yielded SN 1937C, one of the brightest supernova explosions then on record.[44] Due to its visibility and height above Palomar's horizon, Baade diligently recorded its exponentially decreasing luminosity and changing spectrum for 640 days. He was proud of this achievement and spread the word far and wide. Nearly two decades later, four physicists and astrophysicists, trying to understand the creation of the known elements in the universe, would exhume Baade's records of the supernova light curve and use them to move past a significant hurdle.

The research community worldwide had become electrified by the events taking place in postwar Pasadena and at Palomar, where astrophysics flourished. The area became a magnet for visiting scholars who were eager to participate in the intellectual excitement that blossomed during the 1950s. Weekly colloquia, started in 1948 and continuing to this day, were attracting locals and international visitors from all astrophysical disciplines. Astronomers and physicists were flocking to the home of Palomar director Ira Bowen for weekly beer, pretzels, and shop talk. Friday evening seminars on nuclear astrophysics, offered by experimental nuclear physicist Willy Fowler, were often followed by alcohol-infused parties and lively discourse.

Among those in attendance was Fred Hoyle—the shy English theoretician who made himself immortal by predicting a nuclear resonance in the triple-alpha process that powers red giants. This resonance relieved a theoretical bottleneck that had cramped the understanding of chemical synthesis in massive stars.[45] He traveled to Caltech's Kellogg Radiation Laboratory to seek confirmation of his theories and greatly influenced Fowler's experimental work, which soon confirmed the existence of the predicted nuclear resonance. Hoyle was not alone in migrating to Caltech and Palomar. Fellow British astronomer E. Margaret Burbidge and her husband, astrophysicist Geoffrey Burbidge, did so as well, both intrigued by Merrill's discovery of technetium and by the local work on supernovae and mass loss in stars. The Burbidges would act as the glue between the Caltech physicists and the astronomers at Santa Barbara Street.

Pasadena became a sweatshop of sorts as astronomers continued to acquire spectroscopic data on stellar compositions, mass loss, and supernovae to help constrain the experimental nuclear reaction rates pouring out of Kellogg Radiation Laboratory. The laboratory, directed by Fowler, measured transition probabilities and other atomic data needed for stellar analyses, paying attention to feedback from astronomers. Theorists then based their work on new data acquired at the telescopes and from accelerator experiments. Later reminiscing about how his field was enlivened and stimulated by astronomical discoveries, Fowler, a Nobel laureate, felt that nuclear physicists had to build a firm empirical basis for stellar nucleosynthesis and stellar instability. He added: "We are working hard at these problems and are having a wonderful time as our five electrostatic accelerators turn out the results."[46]

"The stars are the seat of origin of the elements" became the mantra of Hoyle, Fowler, and the Burbidges as they hunkered down in a windowless, blackboard-studded office in Kellogg, each with a part of the big picture in their pocket. It was understood that all of the hydrogen, most of the helium, and trace amounts of lithium in the universe were forged in the big bang. To synthesize the heavier elements, the team proposed eight distinct nuclear processes that involved the direct fusion of nuclei or the capture by nuclei of protons and neutrons. For a year and a half, they hashed out the complex nuclear processes needed to create each element and its isotopes, and defined the appropriate stellar conditions.

The team hypothesized that abundant elements such as carbon, nitrogen, oxygen, silicon, up to and including iron are created during the brief lives of massive stars. Once nuclear fusion ignites in a star's hot core, it migrates outward to huff in successive shells of fresh hydrogen fuel, generating energy that flows outward and gives the star its luminosity. Synthesizing these elements during a star's evolution along the red giant branch is especially important for the enrichment of the interstellar medium. As Deutsch had discovered, red giants and supergiants lose their outer layers to the interstellar medium in a continuous stellar wind. Laden with carbon, oxygen, and other heavy elements, these winds, in turn, enrich the gas from which new generations of stars form.

Fusing iron into heavier elements draws energy that a star cannot provide. So synthesizing these heavier elements—a majority in the periodic table—requires a different stellar environment. Heavier elements are built up by the capture of free neutrons during atomic collisions. In the process known as *slow* neutron capture, a nucleus is bombarded with a neutron every 100 to 100,000 years. This requires an environment with a density of 1 million (10^6) free neutrons per cubic centimeter. Merrill's discovery of technetium provided the clue for the cosmic location of this process: Such densities are reached in the cores of red giant stars of up to 10 solar masses during the late stages of evolution. About half of the elements beyond iron are formed in this way.

Producing the remaining half of the elements requires yet a different process. To build up a mass of radioactive nuclei, the ratio of neutron bombardment has to be faster than the time it takes for a nucleus to decay into another element. This means one neutron bombardment every 0.01 to 10 seconds. Known as *rapid* neutron capture, it requires a stellar environment with a density of about one hundred quintillion quintillion (10^{38}) neutrons per cubic centimeter, which is one hundred trillion grams per cubic centimeter—nearly a billion times denser even than a white dwarf.

Californium and the Bikini Test

While searching for such an astrophysical environment, the team became aware of a newly declassified hydrogen bomb test that had taken place in November 1952 on Bikini Atoll. During the explosion, the bombardment of uranium by an instantaneous flux of neutrons synthesized californium-254. Visible in the debris of the explosion, californium-254 is an artificial radioactive isotope with a half-life for spontaneous fission of about 60 days. The example of bombs creating heavy elements on

Earth through radioactive decay provided the team with a clue to the formation of heavy elements in stars. But what kind of astronomical monstrosity would act like a hydrogen bomb? As the team pondered this question, one of them recalled the light curve of Baade's supernova 1937C.

The similarities between these two catastrophic events, the hydrogen bomb and the type I supernova—the decay times of the radioactive isotope in the bomb and the light intensity of the supernova, the extreme temperatures and pressures, bursts of neutrons, and glowing ejecta—set the astrophysicists' minds churning and modeling. The Burbidges, Fowler, and Hoyle began collaborating with Baade and with theoretical physicist Robert Christy, who had worked on the Manhattan Project. The similarities suggested to them that the same radioactive isotope was present in both the debris of the bomb and the type I supernova. Thus, the team speculated, supernovae must develop neutron densities high enough to produce heavy elements. They were certain they had found their "astronomical monstrosity" and published their results in 1956.[47] They wrote that the production of californium-254 in a terrestrial hydrogen bomb provided evidence for rapid neutron capture, just as the production of technetium in the cores of red giant stars provided evidence for slow neutron capture. Although their inference was correct that exponential decay indicates radioactive material had been synthesized in the supernova, it would be shown later that the shape of the light curve was due to the decay of nickel-56 to cobalt-56, then to iron-56. Yet the genius of these authors was to argue that these processes could synthesize heavy elements in stars *on a cosmic scale*.

Out of this fertile interplay sprang a highly creative review paper with the crisp title "Synthesis of the Elements in Stars."[48] The unusually long, 105-page paper, published in 1957 and often cited simply by the authors' initials B²FH, laid out a comprehensive theory of nucleosynthesis taking place in the cauldrons of stellar interiors. It pointed to crucial work done at Palomar on mass loss, chemical abundances, technetium, and supernovae. Its far-ranging conclusions galvanized the astrophysics community and enshrined nucleosynthesis as a topic of study for generations to come.

Since ultimately it was found that californium-254 was not directly detected in the spectra of core-collapse supernovae, the source of the rapid neutron-capture process (*r*-process) material remained hotly debated for over half a century. This all changed on August 17, 2017, with the detection of gravitational waves emanating from a pair of merging neutron stars that had been discovered by the Laser Interferometer Gravitational-Wave Observatory (LIGO). Several teams worldwide raced to identify the event optically in a galaxy many millions of light-years away. The radioactive afterglow from the merging pair was discovered by a nimble team of astronomers at Carnegie Observatories and UC Santa Cruz, working with the 40-inch Henrietta

Swope telescope at Las Campanas Observatory in Chile.[49] They located the afterglow in the nick of time before the object set in the west, took the first spectrum of it, found its redshift, and won the high-stakes race. The event was then intensely followed by astronomers around the world. The way in which the afterglow declined in intensity indicated it was due to a mixture of different r-process elements expanding at the incredible speed of about 60,000 km/s, which is 20% the speed of light. The observations provided strong evidence that many heavy elements—including gold and platinum—are synthesized in the extremely dense matter produced during the collision of two neutron stars. For the first time in history, the nature of such cataclysmic events as mergers of neutron stars could be studied through both electromagnetic and gravitational-wave observations.

Heavy Metal

Considered the father of Caltech astronomy, Jesse Greenstein was charged with creating a graduate department of astronomy in time for the commissioning of the 200-inch telescope. When he joined the faculty in 1947, there was only one other astrophysicist on the faculty, the hellaciously brainy Fritz Zwicky. There was also an alliance with the astronomers of Carnegie's nearby Mount Wilson Observatory. Over subsequent decades, Greenstein would pursue observational astrophysics with every spectroscopic instrument and focus available at the 200-inch telescope. His favorite was the high-dispersion spectrograph at the coudé focus located below the observing floor, as it allowed him to disentangle heavily blended stellar lines. Yet investigating the claims made by B²FH would challenge the spectroscopic capabilities of even the relatively new 200-inch telescope and its coudé.

> **No golden rings:**
> "Another solar system might have condensed out of material consisting mainly of hydrogen and gas ejected from stars which had gone through hydrogen burning, helium burning ... but not the r process. Thus, in such a solar system the inhabitants would have a very different sense of values, since they would have almost no gold and, for their sins, no uranium."
> (E. M. Burbidge et al., "Synthesis of the Elements in Stars," *Reviews of Modern Physics* 29 [1957]: 547–650)

To settle whether interstellar material had undergone enrichment over time, Greenstein, aided by a grant from the Air Force Office of Scientific Research, began a program to measure the chemical abundances of stars spread far and wide in the Milky Way galaxy. He and postdoctoral fellows Lawrence Helfer and George Wallerstein chose to observe four K giant stars representing extremes of chemical enrichment. One star was located in the disk of our Galaxy, one star moved at high velocity through the halo, and two stars belonged to globular clusters far above the galactic plane. Age calculations based on recent theories of stellar evolution, and the color-magnitude diagrams of Allan Sandage and collaborators, informed the trio of astronomers that the star in the disk was young, while the other three stars were nearly as old as our Galaxy. According to B^2FH, the three old stars should be made up of the same primordial material out of which the Galaxy formed. The young star, on the other hand, should be made up of material enriched over billions of years through stellar nucleosynthesis and recycling of material.

The coudé spectrograph had just been commissioned, fortuitously timed for the swiftly developing fields of cosmic abundances, stellar evolution, and nucleosynthesis. It is difficult to imagine today, with fast CCD detectors, modern spectrographs, and automatic guiders, what the astronomers then had to endure to obtain good data. Even with the new spectrograph and the world's largest telescope, it took about 10 hours to expose a spectrogram for each star. With no warm or lighted control room as a retreat, astronomers spent the night in the depths of a dark dome or below the observing floor in the cold, concrete coudé room. They kept the starlight centered on a narrow slit by manually guiding the telescope, and periodically checked the instrumental setup.

For each night's observing, the reward was a single spectrogram recorded on the emulsion of a long, thin photographic glass plate. When the dome closed at dawn, the astronomer secured the plate in a light-tight box and carried it downstairs, around the narrow curving corridors near the perimeter of the dome, to the thick, varnished mahogany door labeled "Darkroom." Again working in total darkness, the sleep-deprived astronomer bathed the plate in developer, swished it in "stopper" solution, and rinsed it with water until a black windup timer buzzed jarringly. The total darkness ended only when a dim yellow light was switched on to reveal whether there was a useful spectrum on the plate. Observing required patience and nerves of steel.

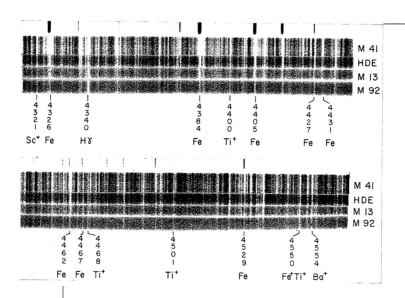

Spectrograms of four K giant stars inhabiting the Milky Way: a star in the open cluster M41 in the disk, HD 232078 in the halo, and one star each in the globular clusters M13 and M92. Whereas the first spectrum, of a disk star, contains hundreds of absorption lines (appearing as vertical light bands), the other three spectra show relatively few absorption lines because the stars are very deficient in heavy elements. These four spectra proved that there is no universal, constant chemical abundance. Rather, heavy elements vary from star to star. Figure 1 from H. L. Helfer, G. Wallerstein, and J. L. Greenstein, "Abundances in Some Population II K Giants," *Astrophysical Journal* 129 (1959): 700. © AAS. Reproduced with permission.

When Greenstein and his two postdocs lined up their four spectrograms, the prize revealed itself after a brief first inspection. As expected, the Balmer line of hydrogen, Hγ at 4,340 angstroms, appeared uniformly strong in all four stars since they have the same spectral type. However, whereas the spectrum of the disk star was crowded with hundreds of absorption lines formed by heavy elements, the spectra of the other three stars showed relatively few such lines. This dearth of lines, it turns out, is due to the fact that halo and globular-cluster stars are deficient in heavy elements by a factor of 10 to 200 relative to the sun and other disk stars! In addition, the absorption bands in the two globular-cluster giant stars revealed for the first time that in those stars carbon is depleted and nitrogen enhanced. In the envelopes of such stars, the ashes of nuclear hydrogen burning are present and visible.

With the publication of their definitive observations in 1959,[50] Helfer, Wallerstein, and Greenstein settled the issue of abundance variations, and the notion of universal cosmic abundances rapidly gave way to the acceptance that stars have a broad range of metallicities. The chemical elements were not *all* produced in the big bang, but mostly in the stars themselves. A star's cache of metals strongly depends on the enrichment history of the gas from which it formed. Ultimately, the cache depends on the deaths of previous generations of stars.

NECROPOLIS

The sun, a middle-aged star born in the disk of the Milky Way 4.6 billion years ago, is a beneficiary of past chemical enrichment—the bequest of stars that died before it was born. In another 5 billion years, the sun will exhaust the fuel in its core, having fused hydrogen to helium, then helium to carbon and oxygen through Hoyle's triple-alpha process. The sun's core temperature will not be high enough to ignite the carbon, the next step in thermonuclear energy generation. Bereft of energy production, an inert mass of carbon and oxygen—the ashes of thermonuclear burning—will build up in the sun's core as it moves from the main sequence to the red giant stage of evolution. The sun will contract and convulse, creating a wind that will carry its tenuous outer layers into space, thereby exposing its extremely dense, hot, luminous core. The ultraviolet radiation emitted by the core will ionize the shells of evicted gas, which will temporarily glow as a colorful "planetary nebula." As the last of the thermonuclear fires in the dying red giant ember burns out, the core will enter the white dwarf stage of evolution. This is the fate shared by most stars between one-half and eight solar masses, the vast majority of stars in the universe: they end up turning into white dwarfs.

The internal physics and structure of white dwarf stars dramatically differ from those of the sun and other stars on the main sequence. Left without thermonuclear fusion reactions to produce an outward flow of radiation, the white dwarf cannot support itself against gravitational contraction. It shrinks in diameter to about the size of Earth, when electron degeneracy—a quantum physics phenomenon that prevents electrons from getting closer to each other—takes over and counterbalances the force of gravity.

Jesse Greenstein was captivated by the physics of highly compressed gases, a state that is impossible to produce in earthly laboratories but exists in white dwarfs. However, observing the faint light of these very small stars to measure their colors, to determine their compositions, and to determine the strength of any magnetic fields challenged even the light-gathering power of the 200-inch telescope. Thus, for a time, Greenstein had nearly a world monopoly on the spectroscopy and photometry of white dwarfs. Due to the paucity of available data to classify their properties, Greenstein teamed up with colleague Olin Eggen in the 1950s to compile the first large homogeneous sample of white dwarfs. Greenstein developed a classification system for these white dwarfs, and for decades his and Eggen's work paved the way for further investigations of stars that lie below the main sequence.

CHAPTER 3

Testing Einstein's Theory

As part of his theory of general relativity, Albert Einstein had predicted that objects of high density should display shifted spectral lines. Because of their extremely high densities, white dwarfs seemed to offer lush testing grounds for measuring such gravitational redshifts. A typical white dwarf core has 200,000 times the mass of Earth. Therefore, as it is about the size of Earth, its surface gravity is 200,000 times stronger than Earth's surface gravity. Because of this, white dwarfs are expected to display Einsteinian gravitational redshifts. Greenstein and graduate student Virginia Trimble wrote a paper in 1967 about their pioneering efforts to measure these line shifts in white dwarfs, being candid about the unusual technical challenges they faced.

For example, their ability to make significant measurements was severely limited by the faintness of the stars, by the scarcity and breadth of lines in their spectra, and by complications arising from the state of degeneracy in such tiny stars. The statistics provided by Greenstein and Trimble in their catalog of radial velocities, space motions, radii, and temperatures confirmed the extreme difficulties they faced. Although they measured the velocities of 94 white dwarfs, they deemed only 53 of their measurements reliable.[51] Trimble recounted the near-trauma of the complex decision-making, while centering the measuring engine's crosshairs on a diffuse spectral line, as a balancing act between "setting on the core, the nearer parts of the wing, or on a grain irregularity" in the photographic emulsion. Yet in the end, by distinguishing the gravitational redshift from the Doppler redshift, they established the reality of Einstein's gravitational redshift in white dwarfs and even measured the gravitational potential of the surface layers forming the hydrogen lines.

A further confirmation of general relativity—and the compactness of white dwarfs—came from the first high-quality spectra of Sirius B, the relatively faint white dwarf companion to the famous star Sirius (also known as Sirius A). Although Sirius B is of roughly solar mass, its radius is smaller than Earth's and its volume is less than one-millionth of the sun's. This extreme compactness generates a very high surface gravity, predicted to significantly shift the white dwarf's spectral lines toward longer wavelengths. Taking spectra to test the prediction was tricky because the companion's light is nearly swamped by the glare of the 10,000-times-brighter Sirius A. Greenstein and colleagues obtained the spectra only after much experimentation and problem-solving, both at the telescope and during data reduction.[52] When they measured the Einstein redshift of Sirius B, they again verified one of the key predictions of Einstein's theory of general relativity: photons lose energy as they escape from strong gravitational fields.

> **CRYSTAL STARS:**
> In the 1960s, Jesse Greenstein undertook excruciating efforts to find only 200 white dwarfs. The European Space Agency's Gaia satellite, launched in 2013, has already identified more than 200,000 of them. With this bounty of new data, researchers recently confirmed an extraordinary theoretical prediction made half a century ago. Although a young white dwarf's white-hot carbon-oxygen core material behaves like liquid metal, after billions of years of cooling, the crystallization temperature of this material is reached. This triggers the core to gradually freeze into a crystalline solid from the center outward. The oldest white dwarfs, of which there are billions in the Milky Way galaxy, are likely to be 80–90% crystallized. (P.-E. Tremblay et al., "Core Crystallization and Pile-Up in the Cooling Sequence of Evolving White Dwarfs," Nature 565 [2019]: 202–205.) In about 10 billion years, the sun itself will begin evolving into a single-crystal star, a sugar-cube-sized chunk weighing about 2½ tons. Thus, the sun will die "not with a bang, but a whimper."

Although there is still some controversy, a general working hypothesis is that stellar evolution ends with a black hole for stars more massive than about 35 solar masses at birth, with a neutron star for stars of about 8–35 solar masses, and with a carbon-oxygen white dwarf for stars of approximately ½ to 8 solar masses. Occasionally, however, stellar evolution ends with zilch when a white dwarf in a binary star system becomes overloaded by mass transfer from a red giant companion—or by a merger with its white dwarf companion—and blows itself to smithereens as a supernova of Type Ia.

The prime focus chair and instrument pedestal. Photo by the author.

· 4 ·
Milky Way Archaeology

Chapter 4

How did our Milky Way galaxy form? In a single event—the collapse of a massive cloud of gas and dust that fragmented into billions of stars? Or was its formation an ongoing process in which smaller satellite objects were captured and accreted over time? The discovery of streams of stars looping like strands of spaghetti around our Galaxy's halo has increasingly tilted astronomers' opinions in favor of a prolonged assembly process. It follows that some stars near to our sun must be aliens from former galaxies accreted long ago. This view of galaxy NGC 5907—a very deep image showing the wrapped-around debris of a decaying satellite galaxy—resembles what an observer looking at the Milky Way galaxy edge-on might see from afar. Mapping the stellar orbits of such debris helps astronomers measure the mass of our Galaxy and its dark matter halo. © 2019 R. Jay GaBany, Cosmotography.com

On a clear dark summer night, the Milky Way arches across the sky in glorious serenity. Its motionless appearance hides a more vibrant nature. We know that some of its stars race by our solar neighborhood at odd angles to the galactic plane, and even occasionally in a retrograde motion. Puzzlingly, some stars are rich in heavy elements while others are anemic. As interstellar gas is continually processed, enriched, and recycled, an old star's trash becomes a newborn star's treasure. Recently, evidence has grown that occasionally the Milky Way strips its satellite galaxies of their stars and gas and claims the plunder as its own. What we observe from Earth is an intricate tapestry of matter and energy—both dark and luminous—woven over eons and through vast regions of space: a cosmic palimpsest of sorts.

Shooting Down the Dogma

The Milky Way's age and emergence from primordial gas have intrigued observers from the very beginning. Yet before astronomers began to model its formation, an old dogma held by spectroscopists in the 1930s and 1940s first had to be overturned. The dogma was that there is a single, universal curve of elemental abundances. That is, all stars were thought to share identical proportions of hydrogen and other elements, which the spectroscopists measured from the sun, a few stars, and meteorites found on Earth. Any apparent deviations in a star's chemical composition were attributed to differing physical conditions in its atmosphere. This dogma toppled surprisingly quickly in the 1950s. First, in 1951, Joseph Chamberlain and Lawrence Aller, two guest investigators at Mount Wilson Observatory, found definitive spectroscopic evidence that the metal abundance of two subdwarf stars was only one-hundredth that of the sun.[53] Then, in 1954, Nancy Grace Roman, a young astronomer at Yerkes Observatory, discovered a way to spectroscopically distinguish stars with varying metal abundances. And in 1957 at Palomar, Paul Merrill detected a short-lived radioactive isotope in a star's atmosphere, clinching the idea that nucleosynthesis was taking place deep within the stars themselves. None of these spectroscopic anomalies could be attributed to physical conditions in stellar atmospheres.

In astronomers' parlance, all chemical elements heavier than helium are grouped under the term "metals." Metal absorption lines disproportionately occur in the ultraviolet region of the spectrum and dim the ultraviolet light emitted by a star. Thus, metal-rich stars appear redder than stars of the same spectral type that are deficient in metals, and which don't suffer from this ultraviolet "blanketing" effect.

Photometrically measured colors of a star can therefore be a good indicator of the proportion of metals in the star's atmosphere.

While compiling a general catalog of about 500 nearby high-velocity stars,[54] Roman noticed that a small group of them had peculiar spectral features and colors. Based on their blue-yellow color and the strength of their hydrogen absorption lines, these 17 stars should have been classified as F-type main-sequence stars—except they were far too bright in the ultraviolet. She termed this characteristic an "ultraviolet color excess." Furthermore, their high space velocities and extremely eccentric orbits resembled those of Milky Way halo stars, not of stars in the disk.[55] The stars' other observables, such as colors and magnitudes, resembled those of main-sequence stars in globular clusters. Roman suggested that it would be "extremely interesting" to photometrically measure some globular-cluster dwarfs in three colors to check for any ultraviolet color excess.

The realization that metal abundances could vary from one population of stars to another strongly influenced the subsequent direction of research in Pasadena and at Palomar. For one thing, Roman's catalog of "subdwarfs," as her stars soon were called, spurred further searches that dominated bright-time observing. It motivated the extensive program led by Jesse Greenstein and his team of postdoctoral fellows to analyze the chemical abundances of large numbers of stars. And although studying the kinematics of high-velocity stars and their relationship to our Galaxy's halo had long been a tradition at Santa Barbara Street and Mount Wilson, Roman's discovery that these stars possess an ultraviolet excess disrupted a paradigm and sparked new interest.

The newfound correlations between velocities and metallicities of stars exhilarated Allan Sandage, who yearned for clues to the ancient past of our Galaxy. He considered Roman's work on ultraviolet excess especially significant for theories of the origin of the universe.[56] Often wearing his symbolically villainous black cowboy hat, Sandage employed the 200-inch telescope primarily for difficult cosmological observations. However, if thin clouds or poor "seeing" disturbed the view, he would switch to dependable spectroscopy of nearby brighter objects. One of his standard backup programs was to search for subdwarf stars. He boasted of changing the telescope's optical configuration from prime focus to coudé focus "in 20 minutes flat" to "bang off radial velocities" for stars on his list of candidates. Now, given the sudden rush to measure ultraviolet excesses at Palomar, Sandage came up with a brute-force method to sort between metal-rich and metal-poor stars. By capitalizing on Roman's stars appearing too blue for their spectral type, he figured out a shortcut to avoid her time-intensive method of taking spectra. Since the goal was to recognize ultraviolet *excess*, Sandage reasoned that spectra were unnecessary. Why not just compare a star's ultraviolet, blue, and yellow light? If a star appears relatively bright in the ultraviolet, it has an ultraviolet excess.

Newly intrigued by Roman's work, astronomer Olin Eggen was in the midst of compiling data on nearby high-velocity stars for two catalogs.[57] A Caltech professor with expertise in photoelectric photometry, he decided to test her inferences. In the late 1950s, he began collaborating with Sandage. The two became lifelong friends as their focus evolved toward disentangling the Galaxy's stellar subsystems. Eggen's catalogs, supplemented with radial velocities acquired by Greenstein with the 200-inch coudé and, on nights with mediocre seeing, by Sandage, would become useful reservoirs of program objects. Eggen himself seemed to always be on the mountain doing photometry.

What happened next would leave Sandage "feeling a little guilty." He secretly believed that Roman already had sufficient data in her catalog of high-velocity stars to discover the same correlations that he and Eggen were about to uncover.[58] She simply hadn't thought of looking for them. By analyzing her data they found that the ultraviolet excess increases with increasing space velocity.[59] The thrilling idea that ultraviolet excess might also correlate with orbital eccentricity led them in 1959 to embark on a multiyear observational and analytical program. Their goal: to obtain orbital parameters of individual stars now passing through the solar neighborhood. They hoped that such correlations might help disentangle the spatial, kinematic, and abundance subsystems of the Milky Way. These subsystems are represented by young main-sequence stars in the disk and old globular clusters, RR Lyrae stars, and metal-poor field subdwarfs in the halo.

When Eggen finished calculating the space motions of a sample of 221 high-velocity stars culled from his catalogs, he and Sandage wanted to see the kinds of orbits these stars follow around the Galaxy. However, converting spatial velocities to orbital eccentricities requires serious dynamical analyses. It was their good fortune that Donald Lynden-Bell, a British astrophysicist and dynamicist, was in residence at the Mount Wilson and Palomar Observatories as a Harkness postdoctoral fellow. Eggen insisted that Lynden-Bell provide him with "nice, simple formulae"—just sines and cosines without elliptical functions. Since there were no electronic computers at the time, Lynden-Bell laboriously calculated by hand the orbital eccentricities, angular momenta, and orbits for stars in various model galaxies. Then, admitting that there were large segments of observational astronomy he knew nothing about—and expecting this to be the end of his involvement in the project—he left the rest to Sandage and Eggen.[60]

Chapter 4

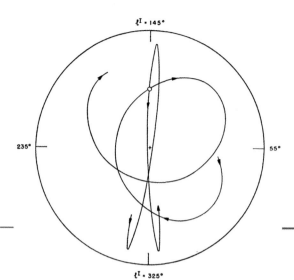

A nearby star's motion through space reveals whether it is an interloper from another part of our Galaxy—perhaps a halo star speeding through the solar neighborhood. Shown here are segments of orbits for two high-velocity program stars calculated by Donald Lynden-Bell. Stars that crisscross the disk of the galaxy experience varying gravitational forces, so their orbits appear more like spirals than ellipses. The two orbits cross paths as they pass through the solar neighborhood, designated by the small open circle near the 145-degree axis. The solar neighborhood lies 10 kiloparsecs from the galactic center, itself represented as a small cross at the center of the outer circle of 20 kiloparsecs radius. Figure 3 from O. J. Eggen, D. Lynden-Bell, and A. R. Sandage, "Evidence from the Motions of Old Stars That the Galaxy Collapsed," *Astrophysical Journal* 136 (1962): 748. © AAS. Reproduced with permission.

The sun fraternizes with stars that belong to the halo as well as with stars that belong to the disk. In a color-magnitude diagram, high-velocity subdwarfs lie slightly below the lower main sequence of normal disk stars. High-velocity subdwarfs are physically, kinematically, and chemically distinct from stars like the sun that occupy the thin rotating disk of the Milky Way. As Eggen plotted his data, the hoped-for revelatory correlations showed up among (i) the chemical composition of the stars, embodied in their ultraviolet excess $\delta(U-B)$; (ii) the eccentricity e of their orbits; and (iii) the maximum height Z_{max} a star reaches above or below the galactic plane during its orbit, related to the velocity $|w|$ of a star in the Z-direction, perpendicular to the galactic plane.

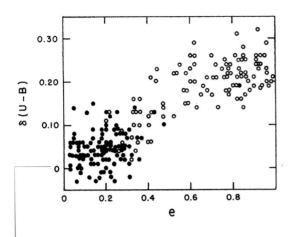
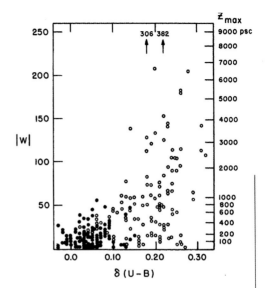

The correlation between eccentricity e and ultraviolet excess $\delta(U-B)$ became a brick in the edifice of Milky Way archaeology. These figures from Eggen, Lynden-Bell, and Sandage demonstrate that low-velocity stars with a small ultraviolet excess hug the plane and follow nearly circular orbits (black dots at lower left in each figure), while high-velocity stars with a large ultraviolet excess follow nearly radial orbits and soar to great heights Z_{max} above and below the galactic plane (open circles at upper right). $|w|$ is the velocity of a star perpendicular to the galactic plane. Figures 4 and 5 from O. J. Eggen, D. Lynden-Bell, and A.R. Sandage, "Evidence from the Motions of Old Stars That the Galaxy Collapsed," *Astrophysical Journal* 136 (1962): 748. © AAS. Reproduced with permission.

In 1955 and 1956 Sandage took up Roman's request for three-color photometry of stars in globular clusters to look for evidence of ultraviolet excesses. Based on his observations in three colors with the 200-inch Hale and 100-inch Hooker telescopes, he found unexpectedly high values of ultraviolet excess for stars in the globular clusters NGC 4147[61] and Messier 3.[62] This meant that field subdwarfs and globular-cluster stars have low metal abundances. Combined with Sandage's earlier work on globular cluster ages, it also meant that these stars are among the oldest in the Milky Way. This finding agreed with the conclusion by Burbidge, Burbidge, Fowler, and Hoyle[63] that interstellar gas is incrementally enriched with heavy elements by each generation of stars over cosmic timescales. Thus, metal-enriched stars are younger, anemic stars are older—and correlations with ultraviolet excess are in effect *correlations with time*. If these correlations were forged during the formation of the Milky Way galaxy, perhaps they could serve as tools for a cosmic archaeologist.

The hope was that differences in stars' kinematic behavior over time might reveal an evolutionary sequence in the dynamic structure of the Galaxy. At this stage, Sandage called Lynden-Bell back to his office and asked, "What does this all *mean*?" Before answering, Lynden-Bell convinced himself that objects that rarely interact with each other over cosmic time are likely to preserve a sort of fossil record of the dynamical conditions under which they were born. Second, the products of nuclear fusion taking place deep within the interior of an individual star rarely mix with its atmosphere, where chemical abundance is actually measured. Thus, any primordial correlations between stars' chemical abundances and orbital characteristics would also be preserved through cosmic time. It follows that a star's current orbital eccentricity is a consequence of the conditions at its birth, even if the star is 10 or 15 billion years old. The immutability of some of the orbital parameters and their correlations allowed Eggen, Lynden-Bell, and Sandage to take the final leap to model the formation of the Galaxy. Listing the many dynamical pillars their theory would have to respect, including the freefall of gas clouds and dissipative flattening of the disk, Sandage quipped, "and that's all there is to it."[64]

According to their model, some 10 billion years ago a huge protogalactic gas cloud 10 times the current diameter of the Milky Way galaxy decoupled from the universal expansion and began to collapse under its own gravity. The cloud was either already slowly rotating, or it acquired spin through torques from similar gas clouds nearby. Condensations that formed during the collapse became globular clusters and halo stars, an idea that agreed with the small spread in their ages that Sandage had deduced from observations. Stars followed elongated orbits that reflected the nearly radial infall of the gas. The entire collapse lasted around 200 million years. Though relatively brief on a cosmological timescale, it was long enough for many generations of massive stars, each living for only 5 to 10 million years, to enrich the infalling gas with heavy elements.

Since there was no balancing force other than gas pressure parallel to its axis of rotation, the rotating spherical cloud became progessively flatter. At some point the collapsing condensations collided and produced an Armageddon of star formation, whose pressure stopped the flattening. The contraction within the plane of the disk stopped when the centrifugal force balanced the gravitational pull—thus defining the Milky Way's current radius. Given the disk's higher gas density, this is where most of the stars in the Milky Way formed and why we see a band of stars arching across the night sky.

At first, the gas followed the same orbits as the stars to which it gave birth, only to separate from them at their nearest approach to the galactic center, known as perigalacticum. Colliding with other gas clouds, the gas lost energy over time and settled into orbits that evolved to become more and more circular and coplanar. Whereas to this day first-generation stars continue their highly eccentric orbits in the now-collapsed and equilibrated Galaxy, later-generation stars follow more nearly circular orbits that reflect the location at which their parent gas clouds had settled. With each new generation of stars, the ultraviolet excess shrank as the gas became cumulatively enriched from their forebears. Newly synthesized heavy elements were dispersed by supernova explosions, red giants generating winds, and planetary nebulae dissipating into the interstellar medium.

While Eggen, Lynden-Bell, and Sandage had specifically modeled the Milky Way galaxy, their picture was considered to be applicable to the evolution of all galaxies. How some galaxies evolved into ellipticals and some into spirals remained a mystery.

Donald Lynden-Bell, Allan Sandage and Olin Eggen (left to right) published their model of how our Galaxy might have formed in 1962. Entitled "Evidence from the Motions of Old Stars That the Galaxy Collapsed," *Astrophysical Journal* 136 (1962): 748, it joined the ranks of transformative papers referenced only with the authors' initials, ELS. The three authors reunited at Greenwich Observatory in 1995, 33 years after publication of their pioneering theory that the Milky Way formed from the monolithic collapse of a huge cloud of gas. Courtesy Gerry Gilmore.

CHAPTER 4

Canaries in a Coal Mine

Interest in the new field of galactic archaeology, embodied in the ELS model of a rapidly collapsing protogalactic cloud, was sidelined for several years by the 1963 discovery of quasars. Then, as astronomers worldwide began testing the model's observable predictions, one weakness stood out. All of the observed halo stars had been sampled from a limited region within a few hundred parsecs (around 1,000 light-years) from the sun. Yet the Milky Way's populations of about 200 billion disk stars and at least 10 billion halo stars extend across 30,000 parsecs (100,000 light-years). The sample of stars used for the ELS model was seriously deficient in the most distant stars, those that presage the pitfalls of any dynamical model, in effect acting as "canaries in a coal mine."

Halo stars are among the oldest known objects in the universe, and so hold key information about the Galaxy's early kinematic and chemical history. Yet well into the 1970s, their systemic features were poorly known. ELS had claimed that the metal abundance of halo stars increases from the most distant to those nestled in close to the galactic center. But Carnegie astronomer Leonard Scarle countered their claim acerbically with, "The effect does not seem to have been looked for, and, indeed, the available abundance determinations have been neither numerous enough nor sufficiently trustworthy to permit the discussion of such questions."[65]

Searle and Robert Zinn—a Carnegie postdoctoral fellow and close collaborator—realized that the ELS picture, although wholeheartedly accepted by many astronomers, was very limited in scope. With a detailed plan to acquire high-quality data out to the most remote parts of the halo, they embarked on an observational study expecting to find an abundance gradient that decreased with increasing galactocentric distance. In the spring of 1976, with Bev Oke's hefty, sometimes balky, multichannel spectrometer attached to the Cassegrain focus of the Hale telescope, they began to spend night after night observing individual stars in globular clusters. By targeting red giant stars that are intrinsically bright, they could push the boundaries of their sample out to 30 kiloparsecs from the galactic center—more than 30 times the observational limit of ELS. While ELS had used simple broad-band photometry to estimate abundances, Searle and Zinn measured "line-blocking" in the blue region of low-resolution spectral scans. They then repeated their measurements over several nights to beat down the errors. It was a formidable challenge to navigate the crowded fields of such distant, faint globular clusters. In the end, they measured elemental abundances in 177 stars in 19 globular clusters orbiting in the outer halo of the Milky Way. The size of their "admittedly meager" sample was dictated by the paucity of reliable color-magnitude diagrams from which they could select red giants for their observing list.[66] They increased their sample size to 44 globular clusters by adding reliable data from the literature.

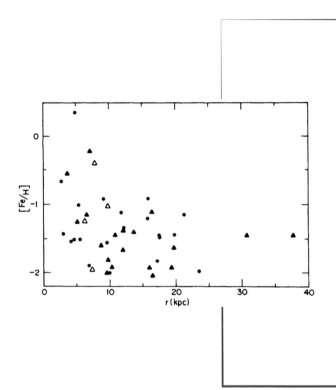

This famous plot summarizes the evidence for an abundance gradient in the outer halo of the Milky Way, as presented by Leonard Searle and Robert Zinn in 1978. A measure of iron abundance, Fe/H, is plotted against r_{gc}, a cluster's distance from the galactic center. The metal abundance of the 44 globular clusters in the sample decreases with increasing distance over the span of the inner halo (0 to 8 kpc), which confirms the ELS model. However, this is in stark contrast to the outer halo (8 to 40 kpc), where heavy-metal abundances appear uncorrelated with galactocentric distance. Figure 9 from L. Searle and R. Zinn, "Compositions of Halo Clusters and the Formation of the Galactic Halo," *Astrophysical Journal* 225 (1978): 357–379, with permission of Robert Zinn.

The way in which the properties of globular clusters, such as chemical abundances, depend on galactocentric distance provides a diagnostic tool for evaluating theories of galaxy formation. Searle and Zinn's new data contradicted the idea that globular clusters formed exclusively during the gravitational collapse phase of our Galaxy. How could clusters even form in the tenuous outer regions of the ELS protogalactic cloud? And, in the ELS model, how could some halo stars and clusters follow retrograde orbits? It appeared that the birth of the Milky Way may not have been smooth and predictable after all.

While formulating their new model, Searle and Zinn's first goal was to sort out the hodgepodge of chemical abundances found in the outer halo in order to make reliable, if broad, inferences about the halo's kinematical behavior and dynamics. What they unearthed so astonished them that they decided to break the mold of old theories and start afresh. Four salient findings guided their new hypothesis:

1. In the inner halo, the abundance of heavy elements in the stars and globular clusters generally increases toward the galactic center. In the outer halo, the abundance is low and varies by an order of magnitude.

2. The age spread of 1 billion years for globular clusters, newly derived from color-magnitude diagrams, was around five times longer than—and thus incompatible with—the 200-million-year collapse time Sandage had deduced for their sample.

3. There was no correlation between abundances and kinematics for globular clusters with abundances below 10% of solar metallicity.

4. Not only were the globular clusters occupying the outer parts of the halo *younger* in the mean than those in the inner parts, they also had a much broader distribution in their ages than the inner ones. Some of the outer globular clusters were as old as the inner ones, but many were younger.

In Searle's and Zinn's opinion, the final important point sealed the fate of ELS.[67] In the collapse scenario, it was hard to understand how clusters could begin forming everywhere at about the same time, but continue much longer in the outskirts than in the inner parts. Also begging for an explanation was the fact that some halo stars and clusters move on retrograde orbits.

The Cannibalistic Milky Way

Though Searle fretted that "it is in the nature of cosmogony that this hypothesis has to be more impressionistic than we should like,"[68] in 1978 he and Zinn published their new formation hypothesis. They proposed that the outer halo of our Galaxy was formed in a more chaotic process than the inner halo. The lack of correlations and gradients in the properties of the outer-halo stars and globular clusters means that these stars did not form during the original collapse of a single cloud of gas. Instead, they likely formed in what Searle and Zinn named "fragments" that surround our Galaxy. Over time, these fragments—today known to be mostly satellite dwarf galaxies—became ensnared one by one in the gravitational pull of the Galaxy. Therefore, in addition to native-born stars and clusters, the outer halo hosts alien material plundered from dwarf galaxies that formed and evolved separately in foreign environments.

Omega Centauri appears to be a globular cluster, but is it? Its stars are a cocktail of ages and chemical abundances. Hence, Omega Centauri is thought to be the remnant nucleus of a former dwarf satellite galaxy that existed billions of years ago. Passing too close to the Milky Way's powerful gravitational field stripped it of its outer stars and led to its capture and ingestion. So it may be that some of the sun's stellar neighbors are aliens from another galaxy. Credit: NASA/JPL-Caltech/NOAO/AURA/NSF.

Since 1978, astronomers have taken advantage of improvements in instrumentation, technology, and data acquisition to more accurately measure metal abundances and velocities of faint halo stars. In doing so, they have confirmed the lack of an abundance gradient for globular clusters and field halo stars out to a distance of 100 kiloparsecs.[69] Puny, severely metal-poor dwarf satellite galaxies—good candidates for "fragments"—have since been found with the Sloan Digital Sky Survey. Some of the most meager dwarf galaxies contain less mass than a globular cluster, yet are embedded in massive dark matter halos. Others show selective enrichment in r-process elements, as defined by B^{2}FH, spewed from merging neutron stars and supernovae of long ago. Perhaps some of these ghostly dwarf galaxies were stripped of their gas when they veered into the Milky Way's corona.

If the outer stellar halo formed by the accretion of dwarf galaxies, then there must be debris left over from these events. The first glyphs of plunder bobbed to the surface in the late 1950s, when Eggen—gifted with an ability to recall minute details of stellar data—noticed that some spatially separated stars not only had similar spectral features, but also moved through the Galaxy with parallel velocities.[70] He tentatively traced these stars to their place of genesis: a since-disrupted cluster. Although his claims were not widely accepted, Eggen continued to collect evidence for what he labeled "moving groups" of stars. Today, moving groups are interpreted as fossils of past accretion events, and their kinematics provides a window on the Milky Way's formation. In an odd twist of fate, Eggen—the first author of ELS—had unwittingly provided evidence for the competing Searle-Zinn model of galaxy formation. However, the full implications of his discovery remained to be excavated by modern galaxy archaeology.

The Spaghetti Factory

Recently, Searle and Zinn's halo accretion hypothesis has been strengthened by the discovery of debris in the form of long streams of stars swooping through and winding around the Milky Way's halo. Some streams seem to be pulled out of dwarf satellite galaxies, an example being the Sagittarius dwarf currently being accreted by the Milky Way. Some stretch across several hundred thousand light-years, encircle the entire Galaxy, and contain more than 100 million stars. Besides shredding and cannibalizing satellite galaxies, the Milky Way also disrupts some of its native globular clusters.

One such cluster in the process of being tidally stripped, named Palomar 5, was discovered in 1950 by Walter Baade as he examined 48-inch Schmidt plates. At that time,

Palomar 5 was thought to be an intergalactic object and not dynamically associated with the Milky Way. In 1977, Sandage verified from 200-inch plates that it is a sparse but bona fide globular cluster located 23 kiloparsecs from the sun—high above the plane and across the Galaxy. Instead of a typical dense and robust globular cluster containing hundreds of thousands of stars, Palomar 5 is diffuse and extended, with a paltry 10,000 stars. However, it provided the first direct evidence that some globular clusters evolve by losing mass in the form of tidal streams.

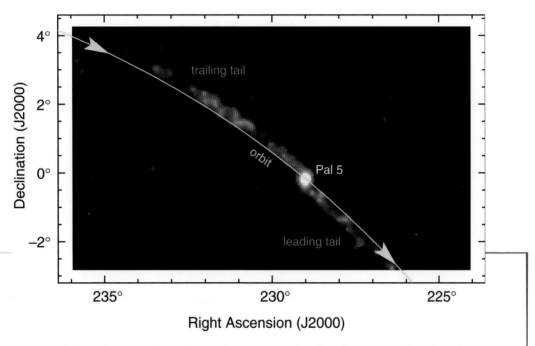

The imbalance between the Milky Way's gravitational pull on the near and far sides of Palomar 5 creates tidal forces that rip out more and more stars from the cluster. These stars then run ahead or trail behind like cookie crumbs. The trails contain more stars than the remains of the cluster itself and span a rather wide 25 degrees (13,000 light-years) across the sky. Models show that the trails were formed by dynamical shocks that occur each time the cluster dives through the Milky Way's disk to the opposite side of its orbit. This is perhaps the last time Palomar 5 will plunge through the disk before its destruction, which is predicted to occur in about 100 million years. Palomar 5's orbit around the Milky Way—shown by the arrows—was reconstructed from the cluster's position, distance, and radial velocity. Credit: M. Odenkirchen et al., "Detection of Massive Tidal Tails around the Globular Cluster Palomar 4 with Sloan Digital Sky Survey Commissioning Data," *Astrophysical Journal* 548 (2001): L165–L169. Image: M. Odenkirchen and E. Grebel.

Chapter 4

In the first three decades of operation of the Palomar 200-inch telescope, investigations produced an approximate age for the Milky Way galaxy, an elegant first theory of the Galaxy's formation by rapid gravitational collapse of a protogalactic gas cloud, and then a radically different theory of formation by the assembly of multiple satellites.[71] While subsequent investigations have strengthened the results from these pioneering works, they have not fundamentally altered them.

The sight of spaghetti-like streams of material being pulled out of disrupted satellites is compelling evidence that the outer Milky Way halo formed through such accretions. Current models show that, over the course of billions of years, most dwarf galaxies leak material that winds around the Milky Way in complex orbits. Some more massive satellite galaxies even merge with the galactic nucleus. In time, all these fragments will be completely assimilated into the Milky Way's bulge and halo. Modeling their orbits provides an estimate of the Galaxy's mass and helps astronomers probe its dark matter halo. Since the Galaxy may have already accreted and cannibalized dozens of dwarfs—and it still is chewing on the 150 globular clusters we observe today—it likely had a much richer assemblage earlier in its history.

It is now clear that the formation of our Galaxy was not a single event in the distant past, but rather an ongoing process of assembly. Many of the globular clusters and smaller satellite galaxies originally orbiting the Milky Way have been cannibalized over the past several billion years. At least three dozen dwarf galaxies, some as small as one-thousandth the size of the Milky Way, have been found in orbit—and since the Milky Way itself appears to be on a collision course with the Andromeda galaxy, the two galaxies themselves will eventually assemble into a bigger galaxy.

The "Phantom" that controlled dome rotation. Photo by the author.

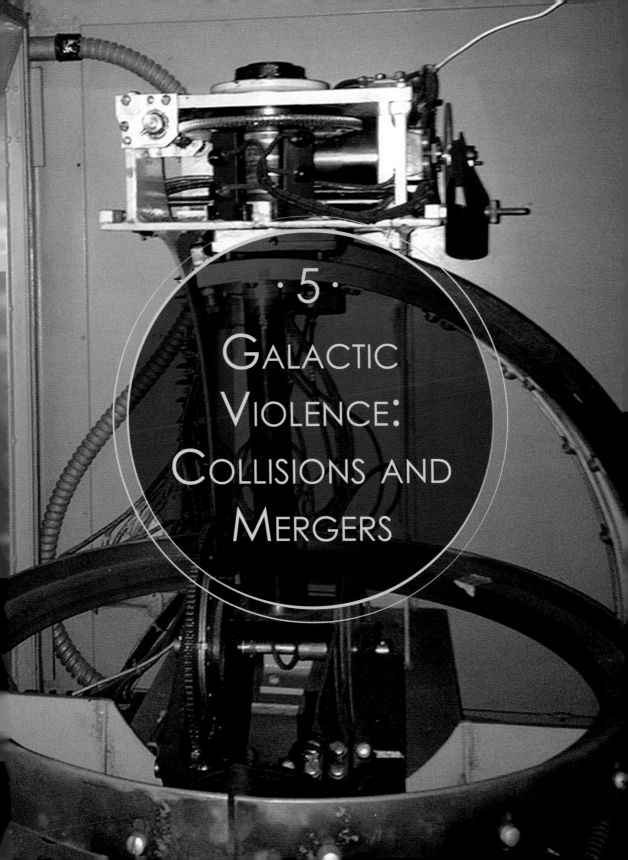

· 5 ·
Galactic Violence: Collisions and Mergers

Chapter 5

Why are the majority of full-grown galaxies in the local universe rotating disks of stars and gas, while a minority of galaxies of similar heft are unstructured heaps of old stars vaguely resembling footballs? It has taken astronomers nearly a century to gather enough forensic evidence to understand this puzzling dichotomy of galaxy shapes. The seeds of understanding were sown by observations of pairs of distorted disk galaxies such as that shown here. 40 million years ago, these two spiral galaxies passed close enough to each other for gravitational tides to develop. In the process, filaments of stars and gas 100,000 light-years long were pulled out of the galaxies. Astronomers now understand that eventually—perhaps a billion years from now—this pair of spirals will probably coalesce and form a single giant elliptical galaxy. Credit: NASA/ESA and The Hubble Heritage Team (STScI).

Blind theorizing is an "empty brain exercise" and therefore a "waste of time,"[72] argued astrophysicist Fritz Zwicky in 1971, three years before his death. His early years, spent grappling with the mysteries of the universe using only the tools of theoretical physics and astrophysics, had convinced him that theory cannot stand alone. It must be supported by observational data. With that in mind, he spent much of his professional life observing, cataloging, and studying the vast contents of the universe. What he saw pitched him against the prevailing beliefs of conservative astronomers and theorists. Yet Zwicky, persevering in the face of their unyielding skepticism, would ultimately help usher astronomers into a radically different view of the universe.

GALACTIC VIOLENCE: COLLISIONS AND MERGERS

COSMIC CHANGELINGS

For centuries, astronomers had assumed that interstellar space was completely empty. This was not unreasonable. Nothing was visible beyond or between individual bright objects, and space appeared to be transparent over large distances. However, Zwicky's free-spirited speculations led him to believe that intergalactic space is instead "sparsely populated by stars, groups of stars, dust, and gases."[73] Until the Schmidt cameras became available at Palomar in 1936 and 1948, the tools he needed to catalog and study the contents of the universe were inadequate. By the mid 1950s—after avidly embracing the art of deep-sky, wide-angle photography—Zwicky began finding thousands of interconnected double and multiple galaxies, groups, and clusters. There were luminous filaments and wisps jutting from some twisted galaxies, taffy-like structures tethered to other galaxies, and even some fanciful ring-shaped galaxies rising out of the shadows. Most of the bridges reaching between galaxies appeared bluish, resembling young stars. None of these strange structures were orderly or predicted, and all were decidedly mysterious.

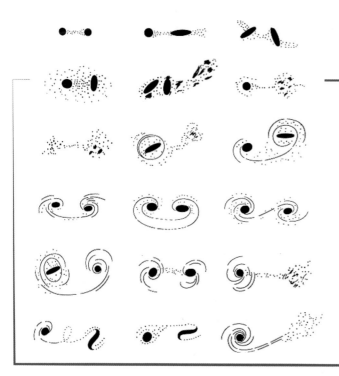

Sketches of peculiar objects that Fritz Zwicky photographed with various telescopes at Palomar. The faint luminous filaments appear to connect to or swirl around neighboring galaxies, suggesting that they are the result of double or multiple encounters between galaxies. Facing the difficulty of photographically reproducing these faint structures for the journals, Zwicky took to sketching them from his photographs. Figure 5 from F. Zwicky, "Multiple Galaxies," *Handbuch der Physik* 53 (1959): 373. Reprinted by permission of Springer.

Chapter 5

Zwicky believed that such complexes held clues to how galaxies and groups of galaxies evolve and were worth investigating in more detail. Besides being a dedicated and careful observer, his strong grounding in physics enabled him to predict findings and interpret most of his observations. Analyzing his photographs from a physicist's point of view, he noticed that many apparently "damaged" galaxies occur in pairs or small groups. The luminous streamers jutting out from opposite sides of such galaxies signaled to him that gravitational forces arising during close encounters must be at play. The galaxies' stars and gas were being pulled out and expelled into space, representing gravitational tides. These tides were analogous to—but much stronger than—the ocean tides raised by the moon on opposite sides of Earth.

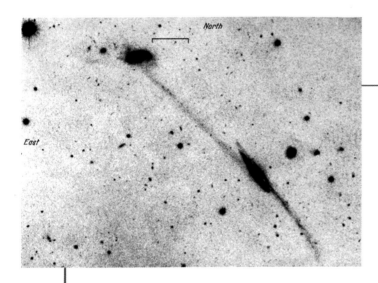

This extraordinarily large system, four times the size of our Milky Way galaxy, was considered by Zwicky to be the most dazzling of the formations discovered in his early sky survey. His 200-inch photograph shows a narrow bridge of stars that stretches across 250,000 light-years to link two galaxies, and a long "tail" that extends to the lower right. Both features are hallmarks of tides produced during a past gravitational interaction. Zwicky speculated that although the two filaments appear narrow, they may be broad curved sheets of luminous material viewed nearly edge on. F. Zwicky, "Multiple Galaxies," *Ergebnisse der exakten Naturwissenschaften* 29 (1956): 344–385. Reprinted by permission of Springer.

In a Darwinesque manner, Zwicky carried his collection of tangled galaxies a step further. He hypothesized that all galactic filaments, no matter how strange their appearance, are caused by tidal interactions. To illustrate his point, he imagined the sequence of events that might have formed an intergalactic bridge between a pair of observed galaxies grazing past each other, one an elliptical and the other a spiral.

Galactic Violence: Collisions and Mergers

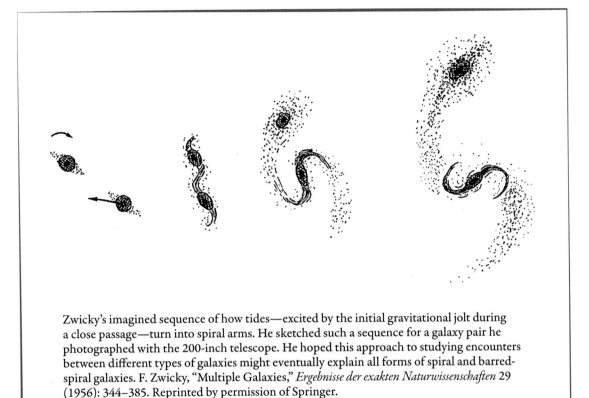

Zwicky's imagined sequence of how tides—excited by the initial gravitational jolt during a close passage—turn into spiral arms. He sketched such a sequence for a galaxy pair he photographed with the 200-inch telescope. He hoped this approach to studying encounters between different types of galaxies might eventually explain all forms of spiral and barred-spiral galaxies. F. Zwicky, "Multiple Galaxies," *Ergebnisse der exakten Naturwissenschaften* 29 (1956): 344–385. Reprinted by permission of Springer.

Endowed with an almost limitless curiosity and deep intuition, Zwicky grasped and predicted astrophysical phenomena with stunning prescience. Yet, during his lifetime, much of his work was widely ignored by the astronomical community. As one example, he wrote that the stellar absorption lines he saw in the spectra of luminous filaments were evidence that the filaments were made of stars. But when he tried to persuade his colleagues that intergalactic space is not empty, he hit a wall of disbelief. The tidal filaments were so faint relative to the night-sky background that most observers could not confirm them with their larger focal-ratio telescopes and less deep exposures. Hence, they were not ready to accept Zwicky's results. His antagonists even claimed that he was recording spectra of moonlight, not filaments of stars. Zwicky—and a few other similarly dexterous souls—had to fight for decades to pry open the eyes and minds of their colleagues.

Chapter 5

To be fair, without a Schmidt camera, astronomers had been constrained to look at the sky piece by piece and object by object. The venerable Harvard Observatory patrol plates had covered much of the sky, but had recorded stars only to about 16th magnitude, not deep enough to detect the faint outer parts of galaxies. Large telescopes, such as Mount Wilson's 60-inch and 100-inch reflectors, had small fields of view (typically less than half the moon's apparent diameter) and had not been specially designed to photograph faint extended features. In addition, due to their long focal lengths and to the low sensitivity of early photographic plates, taking a single deep exposure with these giant reflectors sometimes required more than one night. There were other impediments as well. Some observers neglected to make their telescopes lightproof against stray photons, which fog photographic plates. Dust and uneven metal coatings on the mirror surfaces further scattered and wasted cosmic photons. Finally, the increasing light pollution from nearby cities often obscured the faintest sources of light.

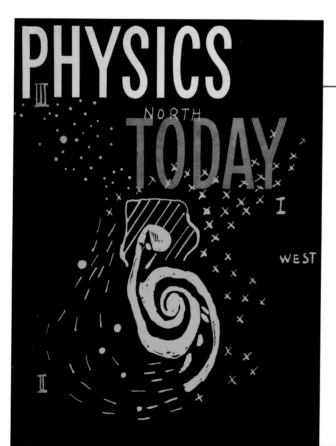

Messier 51, also known as the Whirlpool galaxy, had been photographed and closely viewed innumerable times by competent astronomers. Yet it was the Schmidt camera that first revealed the full network of luminous material splashed around this system. In 1953, Zwicky's sketch of the Whirlpool became cover art for an article in the magazine *Physics Today*. Images and detailed sketches of bizarre galaxies punctuated the text as he laid out his ideas with a high level of sophistication. From F. Zwicky, "Luminous and Dark Formations of Intergalactic Matter," *Physics Today* 6, no. 4 (1953): 7. Reproduced with the permission of the American Institute of Physics.

Galactic Violence: Collisions and Mergers

Zwicky's vision of Messier 51, the famous Whirlpool galaxy, is a case in point. Well known as a classic nearby spiral with a small companion galaxy at the end of its northern arm, M51 (also called NGC 5194/5195) would become a prototype for gravitationally interacting galaxies. North of M51's companion there were luminous "horns," and the southern spiral arm of the main galaxy wrapped around and seemed to reach across the void to its companion. Zwicky correctly interpreted the two conspicuous spiral arms as the "tide and counter-tide" generated in the disk of the main galaxy by the companion during a recent close encounter.[74]

Spectra taken by Mount Wilson astronomer Milton Humason at the prime focus of the 200-inch telescope seemed to show emission lines due to ionized gas in the companion galaxy, and absorption features due to stars in the connecting filaments. From this, Zwicky concluded that the luminous material in galactic bridges and tails consists of stars which are "fairly blue, corresponding in color to the smeared-out spiral arms rather than nuclei of galaxies."[75] In other words, the makeup of stars in the tidal filaments is similar to that of stars in the galactic disks. Therefore, the filaments must consist of disk material.

One of the most fascinating objects in Zwicky's arsenal was the "unique nebula" he photographed in 1941 in the constellation of Sculptor. The object lay so far to the south that the 18-inch Schmidt camera was positioned almost horizontally. The object is a concoction of a faint central galaxy with "spiral star streamers extending outwards to a ring in which many bright stars seem to be concentrated through centrifugal action."[76] By grasping the gravitational nature of the structure, Zwicky demonstrated his deep understanding of physics. Yet how could a galaxy be shaped like a ring, which is gravitationally unstable? At the time, even the brainy astrophysicist could not explain this object.

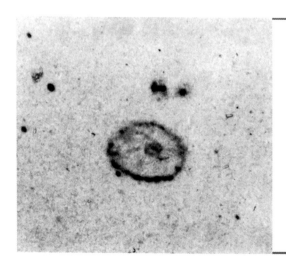

To Zwicky, this object he photographed in 1941 was "one of the most complicated structures awaiting its explanation on the basis of stellar dynamics." An explanation would only be developed decades later, two years after his death in 1974. There is now strong evidence that the ring was produced by a small companion galaxy punching through the main galaxy's disk. From F. Zwicky, "Contributions to Applied Mechanics and Related Subjects," in *Theodore von Kármán Anniversary* (Pasadena: California Institute of Technology, 1941). Courtesy of the Archives, California Institute of Technology.

Chapter 5

By the 1950s Zwicky was largely convinced that most galactic filaments were caused by gravitational interactions and their tides and countertides, even while he occasionally wondered whether some internal stickiness or magnetic process was instead at play in the stringiest of filaments. Having investigated peculiar galaxies for more than two decades, Zwicky made a strong case that observations of close encounters between galaxies could provide insight into the differences in formation that lead to ordinary spirals and to spirals with central bars. In two landmark reviews of 1956 and 1959 he argued that some galaxies can disrupt each other so violently that much of the debris escapes into intergalactic space as interstellar matter.[77] Visible at first as "clouds, filaments and jets of stars," this debris then disperses into the vast space surrounding the galaxies.

Such insights were remarkable at a time when stellar evolution and nucleosynthesis were just beginning to be deciphered and models of galaxy evolution and cosmology were yet to be developed. Zwicky's theories lay fallow for more than a decade, in part because his arguments were not fully quantitative and in part because most astronomers assumed that the distances between galaxies are so vast that galactic encounters are rare and of little consequence. This popular view was bolstered by a series of discoveries made with the 200-inch telescope that doubled—then quadrupled—the scale of the known universe. The quadrupling of scale meant galaxies were four times further apart than previously thought. Already-skeptical astronomers now doubted even more that chance encounters between galaxies could account for the observed number of peculiar galaxies, perhaps paying little attention to the well-documented observation that most galaxies are not isolated in space.

To Zwicky, the stringlike structures found in hundreds of his photographs suggested cohesion. He considered these to be counterexamples to the belief that stellar systems approaching each other with speeds of hundreds of kilometers per second would pass through each other in a head-on collision without disturbance or expelling stars. By 1953, Zwicky understood the gist of gravitational interactions well, yet most astronomers remained oblivious to the power of gravity on cosmic scales. It would take them nearly two decades to catch up.

A Blind Spot

Take as example the case of NGC 5128, a peculiar elliptical galaxy with an intriguing dark gash slung across its midsection. First observed in 1826 by James Dunlop in Sydney, Australia, this galaxy grabbed the attention of Walter Baade and Rudolph

Minkowski when it turned out to be a radio source. Photographs taken with the Hale telescope showed that the gash is a warped band of absorption with some structure, while 48-inch Schmidt photographs showed that the outer parts of the warped band of absorption contain stellar populations similar to those in spirals. The warp told the two astronomers that gravitational tides were at play and that the system was not just an optical superposition. Baade and Minkowski then suggested that a second galaxy, a dust-rich spiral seen nearly edge on, might be interacting with the elliptical behind it while blocking its light.[78]

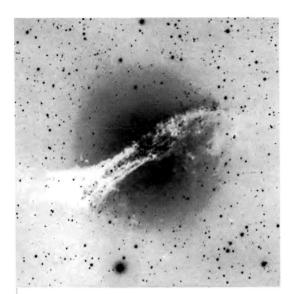 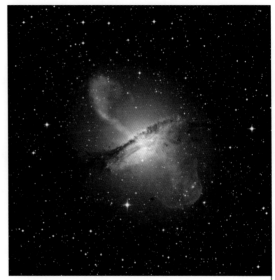

Lyman Spitzer and Walter Baade interpreted this 200-inch photograph (left) of NGC 5128 (the radio source Centaurus A) as a spiral galaxy interacting with an elliptical galaxy. We now know that the entire entity is the remains of a merger between two galaxies around 12 million light-years away. At least one of the galaxies contained gas and dust, which then settled into a disk. The disk is now rotating within the merger remnant, itself a giant elliptical galaxy with the most active nucleus in the local part of the universe. Jets of particles and paired radio-emitting lobes launched by the central black hole show up in the composite image (right). Each radio lobe stretches across 1 million light-years, covering 10 degrees of the sky (20 times the moon's apparent size and well beyond the edges of the color image). Left: Figure 11 from W. Baade and R. Minkowski, "On the Identification of Radio Sources," *Astrophysical Journal* 119 (1954): 215. © AAS. Reproduced with permission. Right: ESO/WFI (optical); MPIfR/ESO/APEX/A; Weiss et al. (submillimeter); NASA/CXC/CfA/R.Kraft et al. (X-ray).

To help him understand the dynamics of this strange system, Baade teamed up with Princeton theorist Lyman Spitzer Jr., an expert on the dynamics of stellar systems. Together, they correctly calculated that as galaxies pass through each other, individual stars are unlikely to collide because their size is negligible compared to their average separations. However, Baade and Spitzer locked horns with Zwicky when they asserted in a 1951 publication that any changes in the velocities and positions of the stars in NGC 5128 would be too small to visibly alter the galaxies' *shapes*.[79] Moreover, they claimed that collisions could not sweep individual stars or interstellar gas out of galaxies. Baade and Spitzer concluded that NGC 5128 might be a very rare case of a spiral disk galaxy caught in the act of sailing through an elliptical galaxy and out the other side *unscathed*. They wrote, "As far as the stars of the colliding systems are concerned, such a collision is an absolutely harmless affair."

Such claims clearly irritated Zwicky, who warned publicly that the two astronomers' views on the dynamical interpretation of multiple galaxies were radically wrong. At the time, Baade was a famous observational astronomer on the West Coast and Spitzer a well-known theoretician on the East Coast. Both had high visibility relative to Zwicky, and their ideas garnered more attention. It is astonishing that Spitzer, an expert in dynamics, was blind to the importance of gravitational interactions and dynamical friction on the system as a whole. In reality, the relative motion of two galaxies generates a changing gravitational field that can significantly deflect the orbits of stars and visibly shred the outer parts of the disks. Zwicky even guessed that encounters "may be so close or even head-on that either considerable disruption of both systems results, or perhaps even total mutual capture."[80]

As the true nature of such galaxies would not be understood for quite some time, a variety of mechanisms other than gravitational tides were invoked during the 1950s and 1960s to explain their peculiarities. These mechanisms included "new physics," magnetic fields, repulsive forces, and even the ejection of entire small galaxies from bigger galaxies. The discovery of quasars in 1963 stimulated ideas about cataclysmic explosions in the centers of galaxies. Cambridge astronomers Geoffrey and Margaret Burbidge, together with theoretician Fred Hoyle, interpreted some faint streamers as having a "tubular form"—resembling a kind of birth canal—for galaxies forming and condensing from the intergalactic medium. Feeling challenged by this, Allan Sandage used photoelectric and photographic observations made at the Hale telescope to show that any claims of recent galaxy birthing were untenable.[81] His measurements of the colors of nuclei in supposedly young galaxies clearly indicated old stellar populations, and his measurements and interpretation have stood the test of time.

Nuggets from a Survey

Things began to change when the National Geographic Society joined Palomar Observatory to produce the Palomar Observatory Sky Survey (POSS), a collection of 1,758 photographs of the sky taken between 1949 and 1956 with the new 48-inch Schmidt camera. The purpose of the POSS was to map the sky in order to ferret out interesting targets for the 200-inch telescope. With many eyes scanning the plates, including Zwicky's, the POSS yielded a trove of new objects. The collection of "peculiar galaxies" grew from a few dozen to several thousand.

One of the first astronomers to systematically search the POSS for peculiar galaxies was Boris Vorontsov-Velyaminov of the Sternberg Astronomical Institute in Moscow. In 1959, he published an *Atlas and Catalogue of Interacting Galaxies*,[82] containing data on 356 peculiar systems with tails, connecting bridges, "common haze," and/or warped dust lanes. The *Atlas* would soon help excite the interest of one of Palomar's own observers.

Though he had dreamed of studying philosophy from the age of 12, Halton (Chip) Arp decided to first study astronomy at Harvard University "to get the facts straight."[83] At the end of his undergraduate studies, his professors urged him to go to graduate school at Caltech, as "they have the new 200-inch telescope." So, in the fall of 1949, having graduated *summa cum laude* from Harvard, Arp joined Caltech's first graduate program in astronomy. For his thesis project, he wanted to search for galaxies with puzzlingly bright energetic nuclei. However, his advisors Baade and Minkowski looked over his proposal and told him: "No, that wouldn't lead anywhere. You should work on globular clusters." Fortuitously, a decade later Arp would get a second chance to pursue his original thesis interests.

By the early 1960s, Arp had become a staff member at Mount Wilson and Palomar Observatories. According to him, he did not set out to open a new field of astrophysics. At the time Caltech astronomers were excited about quasars, privately passing around the coordinates while jockeying for observing time on the 200-inch telescope. As rivalries developed, astronomers began hiding the coordinates from each other and from outsiders. Arp realized that he wasn't privy to this information, so he came up with an alternative plan. He would try to better understand how galaxies form and evolve by studying the ones that didn't fit into Edwin Hubble's iconic collection of "normal," symmetric galaxies.

Sandage—Arp's friend and colleague—had worked for more than a decade to complete *The Hubble Atlas of Galaxies*,[84] a kind of homage to his former mentor who had died before its publication. Hubble's goal had been to discover and photograph the prototypes of galaxy classes in the hope of finding evolutionary connections between them. The *Hubble Atlas* favored galaxies with well-formed, symmetric shapes and structures.

Those galaxies that didn't fit in were dumped into the "Peculiar" section at the back of the book. Because of their "very limited numbers," Hubble (and Sandage) reasoned that such galaxies could be neglected in this preliminary survey.[85]

> **CHIP ARP *IN SITU*:**
> "I spent several years of my life in the prime-focus cage of the 200-inch telescope, taking high-resolution, limiting exposures of peculiar galaxies. It was not boring, because I thought about my subjects and let my imagination roam freely. However, it was a great responsibility, since the telescope was in high demand and so valuable. To be given time on it, I thought, was an enormous responsibility to do something important and worthwhile. I always had a cardboard box filled with the coordinates of objects from zero hours to 24 hours [of right ascension], so no matter how the weather was, or the seeing, I always had what I felt was an important object to observe. I really liked the idea of being able to reach into the box and pull out something that needed observing, that I had my heart in."
>
> (H. Arp, personal interview with the author, 2009)

Arp, however, was fascinated by these rare, distorted objects that often flaunted luminous streamers. He questioned the traditional linear galaxy classification scheme, which was based on physical shapes that Hubble had described as "smooth amorphous ellipticals" and "flattened spirals with star-studded arms." To Arp, these classical categories appeared to be oversimplifications that ignored the details of star formation, chemical composition, and irregular morphology.

Bucking the common belief that peculiar galaxies—though interesting—were statistically unimportant, Arp argued that studying a large number of them might provide new insights into the kinematics and contents of galaxies in general. His first hurdle was Ira Bowen, then-director of the Mount Wilson and Palomar Observatories. Initially averse to yet another large-format photographic survey, Bowen was won over by Arp's cunning argument that the "magnificent Hale telescope, which is due to your great engineering, needs the big photographs to show off its abilities."[86]

To jump-start his collection of oddballs, Arp combed recently published catalogs, borrowed plates stashed in the file drawers of his colleagues, and inspected the POSS prints. At Palomar, he then systematically photographed each oddball with the Hale telescope and occasionally with the big Schmidt camera. He had expected peculiar galaxies to appear bright in the telescope, yet often only saw a faint smudge of light from his target in the eyepiece. He bravely exposed photographic plates for hours, not knowing how much light he would collect. The dramatic revelation would occur only at the end of the night in the darkroom, after he had developed the photographic plates. Then, after he switched on a light box, fantastic details such as filaments, connections, and outer structures would suddenly appear in the darkened crystals of each photographic plate.

By eye, Arp compared the morphological features of each system, then organized and grouped them empirically. For example, galaxies with luminous jets were grouped together, as were those with concentric rings, loops, tails, bridges, long connecting filaments, or companions at the ends of spiral arms. With uncanny judgment, he characterized the nature of the galaxies' peculiarities, and how they vary among diverse groups of objects. His goal was to clarify the inner workings of galaxies themselves, including the underlying physical processes. He hoped that the twisted shapes and curious filamentary linkages he collected would ultimately lead to a better understanding of the workings of the universe. Especially among young astronomers, Arp's contention—that the peculiarities might enable astronomers to analyze the real nature of galaxies—found wide and sympathetic reception.

A Field Guide to Oddball Galaxies

Since viewing high-quality photographs of interacting galaxies was the privilege of a very few astronomers who had access to large telescopes, Arp generously decided to share his collection with the rest of the world. For a 1966 publication, 338 high-resolution photographs were arranged six to a page and printed on glossy photographic paper. The large pageant of objects in Arp's *Atlas of Peculiar Galaxies*[87] precipitated a keen interest, and astronomers everywhere sought copies. The *Atlas* made peculiar galaxies seem more important and compelling to study.

CHAPTER 5

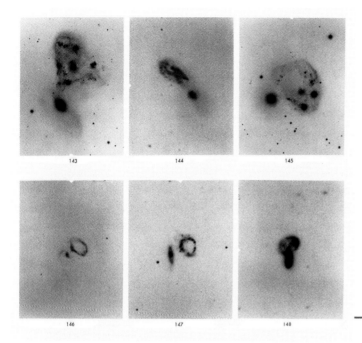

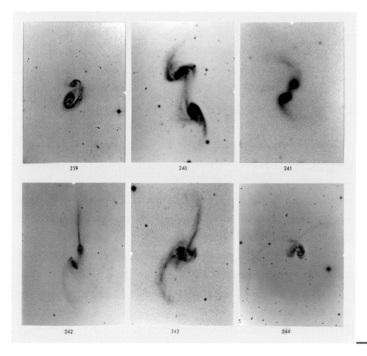

Pages 25 (top) and 41 (bottom) from Arp's *Atlas of Peculiar Galaxies* illustrate his gift for grouping images according to galaxy morphologies. "Some galaxies don't fit and when examined closely, every galaxy is peculiar," he wrote. Explaining the morphologies as the result of gravitational interactions would lay the foundations of our present-day understanding of galaxy evolution. The notches in each image designate north. H. Arp, *Atlas of Peculiar Galaxies* (Pasadena: California Institute of Technology, 1966), with kind permission of Dr. Arp.

The astronomical community was going through a decade of mind-wrenching discoveries: quasars in 1963, the cosmic background radiation in 1965, and pulsars in 1967. Many astronomers had become receptive to new findings that were intellectually overwhelming. Scrutinizing Palomar Observatory Sky Survey plates independently, Zwicky and Arp had each curated a rich menagerie of misshapen galaxies. Counter to Zwicky's intuition that collisions and gravitational tides were responsible for both torn-up galaxies and the beautifully shaped Whirlpool galaxy, Arp continued to believe in unusual gravity and unconventional physics. In his opinion the "so-called canonical collisions" that form galactic bridges and tails require "perfect conditions" that would rarely occur.[88] Instead, and congruent with a school of Russian astrophysicists, he invoked magnetic fields, internal explosions, gaseous hot plasmas, and jets of material shot out of nuclei that tear up galaxies and populate the surrounding regions with streamers. Further embracing the unorthodox view that quasars themselves might be ejected from local galaxies, he veered too far and lost his credibility with many fellow astronomers.

Astronomers' views on interacting galaxies began changing radically in 1969, when the young Alar Toomre—an intellectual and physical giant—arrived at Caltech on sabbatical from the Massachusetts Institute of Technology. A mathematician by training and one of the world's experts on the dynamics of galactic disks, he was looking forward to discussions with Peter Goldreich, Maarten Schmidt, Allan Sandage, and Donald Lynden-Bell, an astrophysicist visiting from Cambridge University, all of whom he considered his "heroes of Southern California astronomy."[89] Yet it was the work by Zwicky and Arp at Palomar that opened the door to the intriguing complexities of peculiar galaxies.

According to Toomre, Arp was an unexpected bonus during his sabbatical. Arp was athletic and charming, with a Clark Gable kind of personality. The two diametrically opposite scientists soon became tennis partners, sparring over more than who hit the better serve. Already cognizant of the zoo of weird objects culled by Zwicky, Toomre's

> "A lot of people went in and feasted on these galaxies and the Toomre brothers' version that they were colliding, and that gravity was what was messing them up. But the Toomres got a restrained warmth from me in return." Asked whether he would do it all over again, Arp laughingly answered: "Yeah, yeah, but I don't think they'd let me!"
> (H. Arp, personal interview with the author, 2009)

interest was further piqued by Arp's *Atlas* of "funny galaxies." Stacks of glossy photographs from the *Atlas* were laid out on long oak tables in Caltech's astrophysics library, where Arp would challenge Toomre to explain the galactic forms one by one. Although he had come to Caltech thinking that Sandage's *Hubble Atlas of Galaxies* was the "next best thing to sliced bread," Toomre began appreciating Arp's alternative *Atlas of Peculiar Galaxies* as equally impressive. It would turn out to have the greater influence on his scientific views.[90]

Toomre was primed for what he called "crazy galaxy shapes," being well aware from previous theoretical work that one galaxy grazing another could produce tides with surprising vehemence.[91] But could even the beautiful shape of the Whirlpool galaxy have been induced by a flyby, as Fritz Zwicky had proposed in 1953? As one of the world's experts on the dynamics of disk galaxies, Toomre was equipped to pursue the question. By teaming up with his younger brother Juri—a specialist in astrophysical fluid dynamics at the Institute for Space Studies in New York—he also gained access to one of the world's fastest computers of that era.

Whereas Arp had to visually sift through many images to study damaged galaxies, the Toomres chose to experiment with model galaxies numerically. Their models were deliberately simple yet efficient caricatures of the stars, dust, and gas orbiting in real galaxies: a central mass point surrounded by a disk of massless orbiting test particles.[92] Over the course of a simulated encounter, two such model galaxies approach each other along parabolic orbits, each feeling a gravitational pull from the other. The test particles of each disk track how the combined gravitational field of the two galaxies varies with time, and how it influences the orbits of stars and hence the appearance of the interacting galaxies. Moving the two central mass points and all test particles computationally forward step by step, the computer produced "snapshots" of the two model galaxies showing how their tidal structures evolve kinematically during and after the encounter. Over hundreds of simulated encounters, the Toomres varied the galaxies' mass ratios, orbital parameters, and spins, as well as the times and directions of viewing the model encounters. Throughout this process, Juri traded visuals and Alar traded computer code back and forth across state lines, sometimes by express mail and other times by Greyhound bus.

Instead of modeling generic, imaginary galaxy collisions, as had several competitors, the Toomre brothers masterfully reconstructed some of the actual shapes displayed by interacting galaxies shown in Arp's *Atlas*. This tactic gave their simulations substantial credibility in the scientific community. Since galactic tides are transient, solely kinematic phenomena, even the strangest, most contorted galaxies don't require stabilization by magnetic fields, gas pressure, or "new physics." Of course, in astrophysics "transient" is a relative term. In an actual galaxy interaction, traces of the damage inflicted by the interaction can last for several hundred million or even a billion years.

GALACTIC VIOLENCE: COLLISIONS AND MERGERS

THE WHIRLPOOL GALAXY AND ITS COMPANION

The Toomres' models of interacting galaxies were not the first, and they were simple by today's standards, yet they racked up an impressive display of contorted beauty. The simulations demonstrated that slowly unfolding near-collisions between galaxies are powerful enough to extend the outer parts of a disk into ribbon-like bridges and tails—or to lay out symmetric spiral arms—simply by gravity acting alone.

Take as example Messier 51—the Whirlpool galaxy sketched by Zwicky and later included as #85 in Arp's *Atlas*. The challenge was to reproduce the interacting pair's morphology: the striking faint streamers around the companion, the freshly excited spiral structure of the main galaxy, and the observed velocity difference between the two galaxies. The Toomres' simulations had shown that encounters between galaxies of unequal mass develop tidal bridges that are strong, narrow, and dense.[93] For their M51 model, they chose a mass ratio of 4:1, with both disks rotating counterclockwise and approaching each other along parabolic orbits.

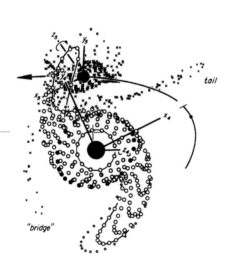

The Whirlpool galaxy and its companion, representing gravitational tides as they developed during the close passage of a large disk galaxy and its quarter-mass companion over a time period of about 100 million years. Left: Figure 21b from A. Toomre and J. Toomre, "Galactic Bridges and Tails," *Astrophysical Journal* 178 (1972): 623–666. Reproduced by permission of Alar Toomre and Juri Toomre. Right: Copyright Jon Christensen.

In this model, on their first close approach the two galaxies experience strong tidal forces that eject material from opposite sides of their disks. As the galaxies begin to interpenetrate, long streams of stars and gas are pulled out into tails, forming filaments of debris that stretch further and further away from the disks. As the galaxies begin to separate, a tide is pulled out from each galaxy's near side, and it evolves into a narrow bridge linking the two. A far-side tide wraps into a fine counterarm that becomes tenuous as it elongates. Stars that are displaced by the encounter will either be captured by the companion or fall back into their "home" galaxy.[94] A recent color image of the Whirlpool galaxy confirms the various streamers, the nearly symmetric strong spiral pattern, and the bridge extending toward the companion that are predicted in the model.

The Antennae Galaxies (NGC 4038/4039)

The Antennae—sometimes also designated Arp 244—are one of the nearest known pairs of merging galaxies. About 200 million years ago the two galaxies passed each other close enough to rouse strong tides. While the tidal tails are now visible as long symmetric filaments resembling an insect's feelers, the bridges are minor and nearly invisible. The Toomres' test simulations had demonstrated that collisions between galaxies of comparable mass create more spectacular tails than do collisions between galaxies of more unequal mass. Such near-equal-mass encounters also inflict sufficient damage to lead to the rapid decay of the orbits and merging of the galaxies. Much stronger forces are evident in the Antennae's tides than in those of the Whirlpool galaxy, which was modeled with a mass ratio of 1:4.

The Toomres were challenged by how to reconstruct the crossing of the tails. The best parameters for the model eluded them, until Juri had the clever idea of changing the viewing angle to above the orbital plane where the action occurs. This allowed them to explore how the model might look from different directions in space, and a new perspective emerged that more closely resembled the geometry of the actual pair.

In this model, during their closest approach the tidal forces on each galaxy's near side pulled out material toward the other disk. Simultaneously, each central mass felt a strong tug toward the other mass and appeared to separate from its own far-side disk material, which was pitched into long counterarms or tails. Recent measurements of the pair's velocity difference and their direction of rotation have confirmed the model's predictions.

> **HIT AND RUN:**
> It is difficult for nontheorists to imagine how extreme and disruptive gravitational tides can become. Familiarity with ocean tides provides little guidance. Tides between galaxies are often much fiercer than ocean tides because (1) relative to their size, galaxies can pass much closer to each other than do Earth and the moon, and (2) tidal forces increase with the third power of the diminishing distance between two bodies. For example, if the moon were moved ten times closer to Earth, typical five-foot-high ocean waves in New York City would become nearly a mile high twice a day. So why do galactic tides often appear as long, drawn-out filaments? One reason is that spiral galaxies are spinning, and another is that the strong tidal forces jolt them over a relatively brief time. A close galaxy-galaxy encounter may itself last only a few tens of millions of years, leading to the tidal jolt. Yet, as the galaxy separation increases following the encounter, the tidally ejected stars and gas keep expanding into luminous filaments that remain visible long after the close interaction.

The Antennae. The left panel is the final snapshot from the simulation of two equal disks undergoing a close encounter. The snapshot shows the model viewed from a direction in space where the tidal tails would appear crossed similarly to those of the Antennae as viewed from Earth, shown in the ground-based image to the right. Left: Figure 23 from A. Toomre and J. Toomre, "Galactic Bridges and Tails," *Astrophysical Journal* 178 (1972): 623–666. Reproduced by permission of Alar Toomre and Juri Toomre. Right: Courtesy François Schweizer.

Chapter 5

Ring Galaxies: Ghostly Apparitions

Tucked into the middle of Arp's *Atlas of Peculiar Galaxies* are three members of a category labeled "galaxies with associated rings." For years, astronomers wondered how such odd, gravitationally unstable ring structures could arise and might evolve. An explanation was missing until 1974, when John Theys, a graduate student at Columbia University, found a first clue. He noticed that the nearest neighboring galaxies appeared to have a "clear preference" for lying along the minor axis of the rings.[95]

One of Arp's large-scale photographs taken with the 200-inch telescope of a ring galaxy named II Herzog 4 provided another clue. Visible adjacent to one obvious ring were traces of a feature that—rephotographed by Roger Lynds at the Kitt Peak National Observatory—turned out to be a second full ring contiguous to the first. And it had a faint nucleus. To Alar Toomre, this was the "smoking gun" of a dual head-on collision between two disk galaxies. "It seemed like a crazy thing to do, bombarding the main disk right along its axis," he recalls of his model simulations, shown below.[96] Amazingly, only head-on collisions would leave the nuclear bulge of a galaxy at the center of its ring. If instead the perturber's aim was off-center, the bulge could be yanked to one side, leaving the resulting ring empty.

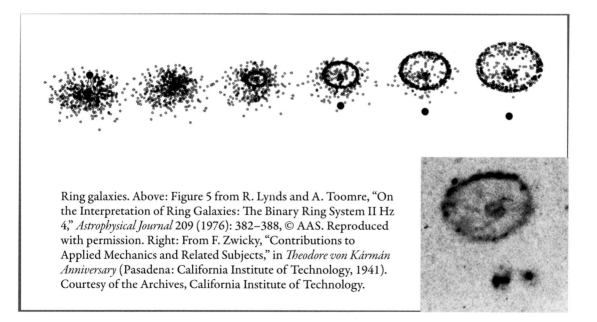

Ring galaxies. Above: Figure 5 from R. Lynds and A. Toomre, "On the Interpretation of Ring Galaxies: The Binary Ring System II Hz 4," *Astrophysical Journal* 209 (1976): 382–388, © AAS. Reproduced with permission. Right: From F. Zwicky, "Contributions to Applied Mechanics and Related Subjects," in *Theodore von Kármán Anniversary* (Pasadena: California Institute of Technology, 1941). Courtesy of the Archives, California Institute of Technology.

Although both phenomena are purely gravitational, ring galaxies differ from tidally distorted galaxies in a subtle way. Alar Toomre demonstrated this with a simple model of Zwicky's "unique nebula," now known as the Cartwheel galaxy. As a compact mass punches through a disk of particles, the gravitational force exerted on these particles increases rapidly. As the disks separate, the gravity in each disk diminishes to its initial value. The resulting rebound alters the orbits of the disk particles, leading to severe radial crowding. The crowding manifests itself as a sharp, ringlike density wave. The wave travels outward through each disk, analogous to a wave formed by a pebble dropped into a pond. The morphology matches the image published by Zwicky in 1941.[97]

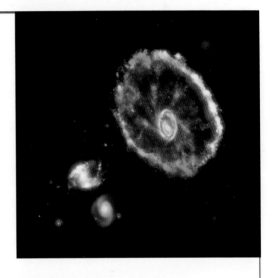

A COSMIC LABORATORY: The Cartwheel galaxy lies about 500 million light-years from Earth. Although its outer ring subtends less than 1/10 the apparent size of the moon, it measures 150,000 light-years across and is large enough to encompass the entire Milky Way galaxy. Around 200 million years ago, one of the two smaller galaxies in the lower left of the image probably collided head-on with what was then a large spiral galaxy. The collision created a density wave that plowed radially outward through the disk, sweeping up and shocking interstellar gas and dust in its path. The shock triggered the birth of millions of massive young stars, which in turn lit up and ionized the surrounding gas to illuminate the crest of the wave. Since stars change in color as they age, strong color gradients are observed in many ring galaxies. The oldest (reddest) stars lie in the central regions of the galaxy and the youngest (bluest) at the peak of the density wave. In a sense, the stages of stellar evolution are mapped across a ring galaxy's face. Hundreds of star-forming regions, all less than 10–20 million years old, are visible as bright blue blobs in the outer ring. The faint arms or "spokes" between the central and outer ring may be relics of the Cartwheel's old spiral structure. This image is composed of four wavelengths: X-rays (purple), visible (green), ultraviolet (blue), and infrared (red). Credit: NASA/JPL-Caltech/P. N. Appleton (SSC/Caltech).

Chapter 5

Forging a New Paradigm

In 1972 Alar and Juri Toomre published their findings in a landmark paper entitled "Galactic Bridges and Tails." Their model simulations not only provided a simple theory of galactic interactions but also opened the larger issue of the important role collisions may play in galaxy evolution. At that time very few astronomers had any idea that dark matter would turn out to be the real boss of these mammoth structures.

Even though strongly interacting galaxies in isolated pairs have traditionally garnered the most attention, gravitational interactions are not limited to such pairs. Galaxies also tend to aggregate in triplets, small groups, and large clusters, some of which harbor thousands of members. As the number of galaxies in a confined space grows, so do the opportunities for interactions. This raises some interesting questions: How much do multiple interactions affect a galaxy's evolution? And how far can forensic science be pushed to identify the precursor galaxies basic to building a complete picture of galaxy assembly?

In large clusters, galaxies tend to whip by each other at speeds up to thousands of kilometers per second. Although this is too fast for them to merge, their collisions can sweep out interstellar gas and produce moderate tidal distortions. In contrast, within small groups of galaxies the relative speeds are lower and more conducive to strong interactions and mergers that produce bridges and tails.

In the mid 1970s, the catalogs and atlases of Zwicky, Vorontsov-Velyaminov, Arp, and others seemed to provide an abundance of interacting galaxies to study. Yet as Paul Hickson, a Caltech graduate student, set out to investigate isolated small groups of galaxies for his thesis project, he found that these collections were unsuitable for statistical studies. Samples chosen on the basis of visible signs of interaction are likely biased in unknown ways, and any derived properties might mainly reflect the adopted selection criteria. As remedy, Hickson collected a new sample consisting of galaxies in small groups. He systematically searched all red POSS plates with a handheld magnifying glass, using strict rules for selection. The resulting catalog, published in 1982,[98] listed one hundred "compact groups" that typically each contain four to six galaxies. The abundance of tidal distortions visible in group members suggested that most of these groups consist of physically associated galaxies, rather than of chance projections.

Hickson's compact groups caught the eye of theoretical astrophysicist Joshua Barnes, then at Princeton's Institute for Advanced Studies, who saw them as promising springboards for studying how galaxies evolve. The landscape had drastically changed since the Toomres' time, when model galaxies consisted of simple point masses and disks of noninteracting test particles. In 1973, theoreticians began pushing the idea that individual disk galaxies, pairs, and groups of galaxies require massive halos of dark matter to stabilize them. Thus, even though Zwicky had introduced the concept of dark matter in the 1930s, and even though the observed rotational motions of stars and gas in Andromeda and in a few other spiral galaxies had supported the idea of dark matter halos, somehow the concept hadn't sunk in until theoretical models required it. By the mid 1980s the consensus among astronomers was that dark matter is a universal phenomenon which—far more than luminous matter—dictates what galaxies do.

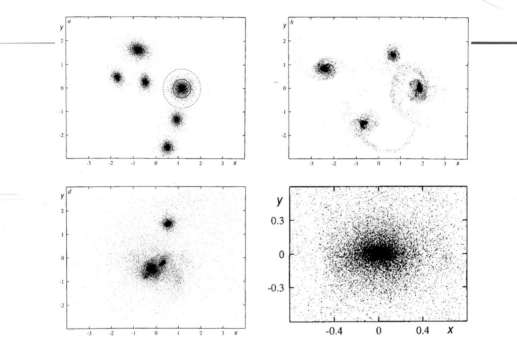

Having published simulations of merging pairs of galaxies in 1988, Joshua Barnes turned to modeling the evolution of a group of six disk galaxies, loosely patterned after Hickson's compact groups. He embedded each galaxy in its own halo of dark matter, marked in the first frame by the solid and dashed circles which enclose 50% and 90%, respectively, of the dark mass. The bulges, disks, and halos of the six galaxies were simulated assuming four times more dark matter than luminous matter, and each galaxy was represented by more than 65,000 particles. The first three frames follow the dynamical evolution of the group as successive pairs of galaxies merge and launch tidal tails that show up in their remnants. The fourth frame depicts the final single merger remnant, a massive "star pile" whose light profile and dynamics look every bit like those of an ordinary elliptical galaxy. Figures 1a, 1b, 1d, and 3 in J. E. Barnes, "Evolution of Compact Groups and the Formation of Elliptical Galaxies," *Nature* 338 (1989): 123–126. Reproduced by permission of J. E. Barnes.

Chapter 5

While we humans cannot see dark matter with our eyes and telescopes, we can detect it through our understanding of physics, along with major innovations in computer and software technology. Barnes ran simulations of interacting galaxies that included, for the first time, massive halos of dark matter surrounding each galaxy. His three-dimensional models also included the visible bulges and disks of galaxies. After carrying out simulations with and without the addition of dark matter, he concluded that dark matter clearly produces better fits to the observations. Since interacting model galaxies require dark matter to produce the observed long tails, real galaxies likely do too. Barnes published his results in 1989[99] and urged others to observe and study Hickson's compact groups.

Although early on the presence of dark matter struck some researchers, including Alar Toomre, as an irritating complication, it turned out to be a blessing in disguise. It fixed several inconsistencies that had plagued simpler models. For example, Barnes's simulations demonstrated that the extra mass in dark matter halos boosts the relative speed with which two galaxies in a group approach each other into the range where it resonates with their internal motions. This boost enables the launch of more substantial tidal tails, resembling those actually observed. And as the galaxies collide, their dark matter halos act as efficient brakes that absorb orbital energy, thus allowing the galaxies to merge.

The Hickson groups being modeled by Barnes caught the eye of famed astronomer Vera Rubin at the Carnegie Institution's Department of Terrestrial Magnetism in Washington, DC. As a mother of four, Rubin had always been determined to pursue a career in astronomy. To insulate herself from questions about her manner or methods of doing science, she pursued topics on the fringe of popularity.[100] So, while everyone else seemed to be prying into galaxy interiors, she chose to work on galaxy exteriors. In 1964, as her career had just begun blossoming, Allan Sandage approached her at an international meeting and suggested she apply for observing time at Palomar. However, the director denied her request for observing time on the 200-inch telescope in 1965, writing: "Due to limited facilities, we cannot accept applications from women." There was only one all-male dormitory on the mountain. It was appropriately named "the Monastery."

Twenty years later, Rubin had confirmed, from the shapes of their rotation curves, that galaxies are embedded in halos of dark matter. With her work widely known, she was finally awarded observing time at the Hale telescope. She showed up on the mountain with two colleagues and with the coordinates of about twenty Hickson compact groups in hand. Due to the faintness of the groups, it took Rubin and her team nearly seven years to obtain images and spectroscopy of some 50 group members. Detailed analysis of the data largely confirmed some of Barnes's predictions: galaxies in close quarters undergo frequent tidal interactions and are unlikely to retain their individualism for long. Eventually, they will all merge into a single, more or less normal elliptical galaxy, finally erasing the tendrils that Zwicky hypothesized and first saw.

Galaxy collisions and mergers are now widely recognized as a significant driver of galaxy evolution. But how do they relate to the more subdued conversion of gas to stars in isolated spiral galaxies? And how do they compare with other dynamical processes such as the stripping of gas and dust from galaxies that fall into clusters? Finally, how do galaxy collisions change with cosmological epoch? Some massive galaxies seem to have assembled very early, at a time when the universe was more dense and galaxy collisions and mergers were much more frequent. In order to detect the evolutionary changes of these galaxies, observers must push to ever higher redshifts. At the present time, astronomers are still struggling with some of these issues, yet recognize that—most likely—a multitude of processes contribute to galaxy evolution.

What has become of early ideas about the evolution of Our galaxy? The Milky Way galaxy is located in a relatively compact group called the Local Group. Astronomers now recognize that our Galaxy must have accreted multiple former companion galaxies over time. If, as predicted, in a few billion years it collides with its major neighbor, the Andromeda galaxy, the two may metamorphose into a single giant pile of stars resembling an elliptical galaxy. The merger remnant NGC 7252 shown here, nicknamed "Atoms for Peace," is such a star pile that resulted from the collision of two large spiral galaxies beginning about 700 million years ago. The process of galaxy assembly is temporarily complete—until a more distant neighbor galaxy comes along for the next potential merger. The remnant, whose two tidal tails extend across about 500,000 light-years in projection, lies at a distance of 220 million light-years from Earth. Credit: ESO.

Chapter 5

Despite these remaining open questions, gravitational interactions and mergers of galaxies have become the new paradigm of galaxy assembly and evolution. It started with Zwicky, who seemed to see what others could not yet, in the forms of luminous filaments and streamers, and who extrapolated them into an evolutionary process. The Toomres' simple model simulations finally won widespread recognition of the importance of galaxy interactions, only for astronomers to find in the last few decades that, once again, there was more than meets the eye: the overwhelming presence of dark matter. Access to better tools enabled these revelations. Yet it was the search for new insights and the dogged questioning of biases by Zwicky, Arp, and others that propelled the story—a story of galaxy assembly, transformation, and evolution.

Shard of Pyrex glass. Photo by the author.

· 6 ·
Quasars: Wolves in Sheep's Clothing

Chapter 6

The beacon of blue light shining from the center of this disturbed galaxy, named Markarian 231, is the telltale sign of a quasar—the highly luminous center of a distant galaxy powered by a supermassive black hole. The luminous material surrounding the quasar is likely a mixture of two galaxies having collided and nearly merged into one. The visible knots and arcs are the result of bursts of violent star formation, the kicking up of high winds, and a feeding frenzy as gas falls into the central black hole. Not only did quasars' extreme luminosities allow astronomers to explore the universe to much greater depths, but explaining their colossal energies led to the modern concept of supermassive black holes. Credit: NASA, ESA, the Hubble Heritage Team (STScI/AURA)-ESA/Hubble Collaboration, and A. Evans (University of Virginia, Charlottesville/NRAO/Stony Brook University).

It all began with a *hiss* from the Milky Way. In 1932 Karl Jansky, a radio engineer at the Bell Telephone Laboratories, was assigned the task of tracking down some annoying static that affected trans-Atlantic radio communications. To his surprise, he found a major source of noise from beyond the solar system in the direction of Sagittarius, the center of our Milky Way galaxy. Unaware of the enormous consequences of his discovery and soon assigned to other tasks, Jansky published his discovery but never followed up.

From Local Hiss to Distant Roar

Fortunately, a young and admirably tenacious American radio engineer named Grote Reber was thoroughly captivated by Jansky's findings. In 1937, at the age of 26, Reber was so eager to understand the fundamental physical processes that generated the celestial radio emission that he designed and built a 50-foot-diameter parabolic radio dish. Operating as the world's only radio astronomer until after World War II, he detected nonthermal emissions from the sun, the plane of the Milky Way, the galactic center, two supernova remnants, and—most importantly—from a source in the constellation of Cygnus. His findings and maps of the radio sky, published in the *Astrophysical Journal* in 1944, would change the course of astrophysics. Neither George Ellery Hale with his vision for the 200-inch telescope, nor all of the astronomers who tried to foresee what they could accomplish with it, could have had an inkling of what revolutionary changes were about to unfold.

At first, progress in the new field of radio astronomy was slow. Then, after World War II, large antennas sprang up in Cambridge and Manchester, England, and in Sydney, Australia. With these antennas, budding radio astronomers surveying the sky discovered amorphous concentrations of radio emission, which they assumed were odd stars in the Milky Way. However, the radio sources could not be identified, nor their nature understood, without photographs of optical counterparts. Thus, the rapid growth of radio astronomy provided an unexpected first challenge for the newly commissioned 200-inch telescope and the wide-field 48-inch Schmidt camera. Well known as consummate observers, astronomers Walter Baade and Rudolph Minkowski were asked to take the deep photographs and spectra necessary to search for the radio sources' optical counterparts. Because of fierce rivalries between the three groups of radio astronomers working out of Cambridge, Manchester, and Sydney, each group secretly sent their list of radio positions to Baade and Minkowski only after securing promises of confidentiality.

Chapter 6

Searching for radio sources was like tapping in the dark. The resolution of any telescope depends on the ratio of its aperture to the wavelength of the observation. The intrinsic angular resolution of the 200-inch telescope at optical wavelengths is 0.01 arcseconds, although the "working" resolution is one arcsecond due to the limitations of thermal "seeing" and the effects of the enclosure. By contrast, the resolution of radio dishes of that era was extremely poor. Although dishes were only a few times larger than the 200-inch mirror, radio wavelengths are thousands to millions of times longer than optical wavelengths. So the positional uncertainties of radio sources were typically half a degree, or as much as the moon's apparent diameter.

During their campaign to identify optical counterparts for the radio sources, Baade and Minkowski sometimes struggled with fields crowded with thousands of stars and faint galaxies. It was akin to searching for a needle in a haystack, with an equally poor success rate. By 1950, only 7 of 67 cataloged radio sources had been tentatively identified optically, and none was solidly confirmed. Subsequent radio surveys—each more technically advanced—were carried out and published, many by the Cambridge University group. Eventually, through the application of ingenious techniques, the various radio groups improved their precision. Hundreds of sources were compiled, and their coordinates were published as the Third Cambridge Catalogue of Radio Sources (Revised), dubbed 3CR.[101] Soon the 3CR became the standard reference for locations of radio sources.

At first, optical astronomers were slow to pay attention to cosmic radio sources. Some sources sat unnoticed for years, until powerful radio emission emanating from a tiny region in the constellation of Cygnus finally caught the attention of the astronomical community. No one yet knew how far away the source was, and no one knew what it was …

Buried within this extremely crowded field rich in foreground stars and glowing gas within the Milky Way, and numerous faint galaxies beyond, lies the optical counterpart to the brightest radio source in the constellation Cygnus, named Cygnus A. But which of the thousands of objects is it? Baade drew a box 18 arcminutes × 10 arcminutes on his 48-inch Schmidt plate to represent the wide range in coordinates provided by rival groups of radio astronomers. Then he found Cygnus A amid a rich cluster of galaxies. It coincided with one of the brightest members of the cluster. Figure 10 from W. Baade and R. Minkowski, "Identification of the Radio Sources in Cassiopeia, Cygnus A, and Puppis A," *Astrophysical Journal* 119 (1954): 206. © AAS. Reproduced with permission.

Cygnus A, earlier noticed by Reber, differed from most other radio sources. Although it is the second-brightest radio object in the sky, Cygnus A had thwarted all attempts to firmly identify it with a celestial object. To improve the radio resolution, in 1951 Graham Smith of Cambridge University linked two radio telescopes together and confined the position of Cygnus A to an error box.[102] As soon as this new position was received at Palomar, Baade photographed the field with the 48-inch Schmidt camera. Since the radio source lies close to the plane of the Milky Way, its error box was congested with stars and galaxies. On his deep photographs, Baade managed to tease out a faint smudge of 17th magnitude located at the position of Cygnus A. The radio emission came from one of the brightest galaxies amidst a distant rich cluster of galaxies.

The next step was the larger scale at the prime focus of the 200-inch telescope. A masterful observer and photographer, Baade had developed a sense for which weather conditions were conducive to slow temperature changes in the mirror, and he took advantage of them for his lengthiest and most difficult observations. Thus, he earned a reputation for extraordinarily sharp and well-focused images on long exposures. On the larger-scale photograph, the smudge appeared bifurcated—separated by a mere sliver of space—and enveloped in a faint oval glow of light. The object seemed to defy classification.

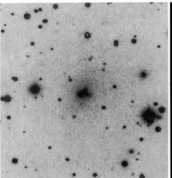 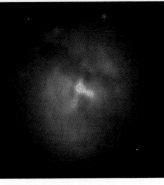

Left: The bifurcated faint smudge appearing on a photograph taken in 1952 was identified as the radio source Cygnus A, and interpreted as two colliding galaxies. The dark streaks are bands of dust. Although it appears faint on optical photographs, Cygnus A is the second-brightest radio source in the entire sky. Right: A modern optical image of the *core* of Cygnus A reveals disorder in the center. What had earlier seemed to be two galaxies is today understood to be a biconical set of bright emission line regions, betraying the fact that a quasar hidden by dust lies at the center of Cygnus A. In 2003, UC Riverside astronomer Gaby Canalizo discovered a small companion galaxy with older stars in the inner region. Cygnus A may be at an advanced stage of two galaxies merging after all. Left: Figure 11b from W. Baade and R. Minkowski, "Identification of the Radio Sources in Cassiopeia, Cygnus A, and Puppis A," *Astrophysical Journal* 119 (1954): 206. © AAS. Reproduced with permission. Right: Keck NIRC2 from G. Canalizo et al., "Adaptive Optics Imaging and Spectroscopy of Cygnus A. I. Evidence for a Minor Merger," *Astrophysical Journal* 597 (2003): 823–831. © AAS. Reproduced with permission.

Minkowski's spectrum of the smudge showed several broad emission lines of highly excited gas, from which he measured a "large" redshift of nearly 17,000 km/s. Cygnus A was thought to be about 100 million light-years from Earth, according to the calibration of the redshift-distance relation at the time. (Today we estimate that it is 800 million light-years away.) This meant that Cygnus A was one of the most distant galaxies observed at the time. Minkowski knew his spectra very well, having spent his life interpreting spectrograms through monocular microscopes. With all this monocular work, the coordination between his two eyes had weakened to the point that he had to perform special corrective exercises. In this case, his experience told him that the emission lines—broadened by rapid chaotic motions of the gas—and the bifurcated image were hallmarks of a rare collision between "two previously separate galaxies" with billions of stars each. They were spiral galaxies, and highly distorted. To be visible at such a distance, whatever was taking place in Cygnus A had to be generating a million times more radio luminosity than is produced by the entire Milky Way galaxy.

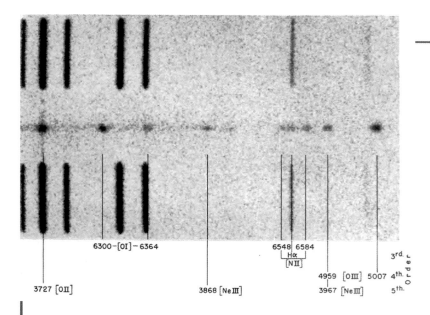

In this grainy photograph, the spectrum of Cygnus A is the clumpy thin horizontal line. It is flanked by vertical laboratory comparison lines. The bright clumps are broad emission features superposed on the barely visible spectral continuum of the galaxy. The existence of "forbidden" lines emitted by singly and doubly ionized oxygen, nitrogen, and neon atoms indicates that Cygnus A contains highly excited and rarefied gas. Rudolph Minkowski took this spectrum with the 200-inch telescope in 1953. Figure 12 from W. Baade and R. Minkowski, "Identification of the Radio Sources in Cassiopeia, Cygnus A, and Puppis A," *Astrophysical Journal* 119 (1954): 206. © AAS. Reproduced with permission.

Before long, radio astronomers discovered that the bulk of the radio emission came not from the optical smudge itself, but from two giant lobes of plasma extending to about 200,000 light-years on either side of the host galaxy. To Baade and Minkowski, the staggering reservoir of energy required to maintain such enormous lobes would have been an intoxicating concept.

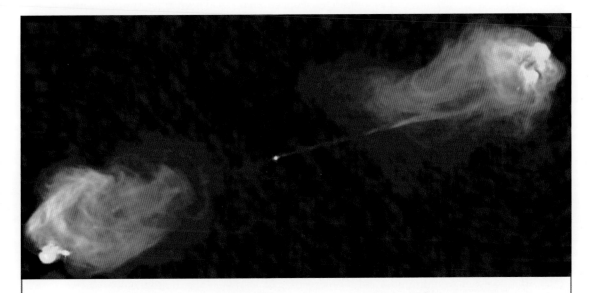

Although the two lobes appear bright in this modern radio image of Cygnus A, they are completely invisible in optical light. Astronomers now know that a supermassive black hole at the center of this disturbed system spews out jets of radiation and particles that inflate the lobes. At the end of the jets, the electrons and protons collide with surrounding gas to form bright "hotspots." Credit: NRAO/AUI.

Baade and Minkowski published their breakthrough study in two now-classic papers, in which they examined the current state of radio sources. They also warned that further identifications would depend on whether radio astronomers could reduce the size of the positional error boxes to well below one degree.[103] The firm identification of Cygnus A with the optical image of a distant galaxy proved that radio sources were not nearby flare stars, as some radio astronomers had suspected. The door to exploring the geometry of the universe through distant radio galaxies was suddenly flung open.

Chapter 6

> "Boy, Baade could make that 200-inch sit up and beg!"
>
> (T. A. Matthews, personal interview with the author, 2013)

With uncanny foresight, Baade and Minkowski understood that the real power of radio-bright sources was their detectability over very large distances. For example, at radio wavelengths Cygnus A is so bright that it could conceivably have been detected by then-existing radio telescopes even if it had been at ten times its actual distance. Not so at visible wavelengths, where it would have been far too faint. Even at only twice its distance, deciphering the morphology of Cygnus A in visible light would have been a struggle, and obtaining its spectrum an impossibility. Clearly, a large volume of the universe easily accessible at radio wavelengths was at the time beyond the reach of even the Palomar giant. More effort and technical innovation would be necessary to push further into that vast new territory.

Again heralding the future, in 1954 Baade and Minkowski made the offhand remark that "a slim chance may exist of obtaining positions of the required high accuracy from observations of the occultation of [radio] sources by the moon."[104] Less than a decade later, such an occultation would thrust the astronomical community into realms that were far beyond anyone's imagination.

As the potential value of radio telescopes became clear, Jesse Greenstein decided that Caltech would build a radio observatory of its own. John Bolton, a well-known English "radio physicist" heading a group in Australia, was hired for that purpose. He arrived in January 1955 and, while scouting for a radio-quiet site on which to build a permanent radio observatory, he installed a temporary 32-foot antenna within sight of the 200-inch dome on Palomar Mountain. Astronomers and electronics specialists then designed a radio interferometer (two separate antennas connected together). Their criterion was sufficiently high angular resolution and sensitivity to resolve sources listed in the Cambridge catalog. In 1959, two steerable 90-foot parabolic dishes were installed in Owens Valley, California, and greatly improved the measured positions of radio sources.

At the time, absorption features in the optical spectra of distant galaxies were very difficult to measure due to the glow of the night-sky spectrum. Even with the light-gathering power of the 200-inch telescope, astronomers had not yet been able to measure redshifts of galaxies beyond $z = 0.2$ (corresponding to a recession velocity of about 60,000 km/s). The lesson learned from Cygnus A was that radio sources offered a promising new way to recognize distant galaxies from their strong emission lines that stand out against the background of the night sky. One such radio source, cataloged as 3C 295, was identified in 1959 by Bolton, who passed its coordinates on to Minkowski.

The radio source appeared ten times smaller than Cygnus A, suggesting that it might be a very distant analogue. On his final observing run at the prime focus of the 200-inch telescope before retirement, Minkowski found a cluster of about 60 very faint galaxies at the location of the radio source. He obtained a spectrum of the brightest galaxy in that cluster. To his delight—and earning a toast from his colleagues—the spectrum boomed with bright lines of forbidden oxygen, from which he calculated a redshift of $z = 0.46$. Assuming that distance is proportional to redshift, the new source was more than eight times further from us than Cygnus A, a new galaxy distance record that was to stand for fifteen years. As we now know, the light reaching Earth today departed Cygnus A about 700 million years ago, while light from 3C 295 departed around 4.6 billion years ago.

Radio Stars?

Enveloped by the fragrance of orange blossoms, Thomas A. Matthews—a freshly minted and newly married Harvard PhD—arrived in Pasadena in 1953 to assume a coveted postdoctoral research position at the Mount Wilson and Palomar Observatories. As he carried his belongings into his assigned office at the headquarters on Santa Barbara Street, he walked into the middle of a memorial service for Edwin Hubble, who had recently died of a massive heart attack.[105] With an education in both optical and radio astronomy, Matthews considered himself a crossbreed. His ambidexterity would enable him to jumpstart what was to become an astrophysical revolution.

By the time he joined Bolton's radio astronomy group at Caltech as a research fellow two years later, Matthews had networked with British and Australian radio groups that regularly provided him with coordinates of radio sources. Examining the characteristics of Cygnus A, he posited that strong, small-diameter radio sources were likely to be distant galaxies. Then, Henry Palmer, "who took the 3CR catalog seriously,"[106] gave hotshot Matthews a list of 30 unresolved sources. Palmer's group had moved three portable cylindrical parabolic antennas to a number of baselines to perform interferometry with the 250-foot radio dish at Jodrell Bank, England. After refining those positions using the Owens Valley interferometer, Matthews began a wholesale effort to identify the optical counterpart of each source. He soon urged Palomar astronomers to become involved in what promised to be a fascinating quest. For one thing, the power radiated at radio wavelengths by these baffling sources tended to be incredibly high. As a first approximation, it was equivalent to a body heated to 10,000,000 K. It was clear that something remarkable was going on.

In 1959, drawn in by Matthews's appeal for collaboration, Allan Sandage began to take photographs with the 200-inch telescope of fields centered on radio sources in Matthews's list. Since a few radio sources had so far been identified with supernova remnants in the Milky Way and with external galaxies, Sandage's attention was piqued by the odd appearance of the first three radio source candidates he photographed. In each case, a starlike object smaller than 1 arcsecond was visible within the 10-arcsecond error box he had drawn directly on the photographic plate. One of them, the radio source named 3C 48, appeared as a 16th magnitude star surrounded by faint "blobby wisps." These wisps were so faint that they became visible on the photographic emulsion only when the plate was viewed nearly edge-on.

These objects looked like radio *stars*, but were they? Sandage realized from his spectrum of 3C 48 that it was a bizarre find. Stars normally radiate as black bodies, with the shape of their spectral continuum depending on the star's temperature. Stellar spectra normally also show absorption lines in their continuum. Yet the spectrum of 3C 48 displayed instead several broad *emission* lines that Sandage could not identify. The spectral characteristics of the object made no sense.

Wildly intrigued, Sandage measured the brightness and color of each starlike object with the photoelectric photometer at the 200-inch telescope. He found that the objects were excessively bright in the ultraviolet, compared to ordinary main-sequence stars. Still more puzzling was that 3C 48's brightness varied significantly from night to night. Sandage was known to be a fanatical observer, so his results were not in doubt: the curious optical-light variations appeared real. But, he reasoned, nothing as large as a galaxy could possibly organize itself to vary so much and so rapidly. So, after his observing run he came down from Palomar and announced, his eyes shining, that 3C 48 was definitely a star. Yet, still mystified, he sat on the data for three years, waiting for an epiphany to clear the logjam.

> "Everybody was so excited because things never before contemplated were breaking at such a rapid rate. The quasars were such an event! Every time someone came down from the mountain, there was a lot of excitement. The spotlight was on, 'What the holy hell are these radio sources that remain unidentified?'"
>
> (A. Sandage, personal interviews with the author, 2007–2010)

ENERGY CRISIS

By sheer accident Maarten Schmidt—a bright young Dutch-born astronomer interested in Milky Way dynamics and star formation—inherited the job of identifying radio sources as Rudolph Minkowski retired in 1960. Ira Bowen, director of Mount Wilson and Palomar Observatories, had convened a group of Minkowski's colleagues to ask whether some of them wouldn't mind taking turns to continue his work as it "seemed to be important." Minkowski was, after all, the gifted spectroscopist who had worked on Cygnus A with Walter Baade. Although Schmidt inherited Minkowski's entire observing program by default, he was not exactly an unwilling participant.

There was growing suspicion that nearby radio galaxies with their giant lobes presented an extraordinary energy dilemma. Evidence had been mounting that a colossal 10^{60} ergs of energetic particles and magnetic fields were involved in the radio source phenomenon. This is roughly equivalent to the total amount of energy emitted by a large galaxy like our own Milky Way—with a luminosity of 10^{43} ergs per second—over the age of the universe. Schmidt thought that was a lot of energy to be stored in a "cloud," or whatever it was, and its source was a mystery to him.

In addition to his existing program, Schmidt began taking spectra of radio source candidates with the 200-inch telescope. Matthews provided the coordinates derived from his own observations at Owens Valley, from his searches of the Palomar Sky Survey prints, from examining photographic plates he asked Sandage to take, and from the British and Australian radio groups. Once in a while, the source in the error box on the finding chart would again resemble a star. Three of these objects, including 3C 48, had already been studied and their spectra were remarkably uncharacteristic of stars. In one case, there were some faint wisps; in another, one strong emission line; and in the third, no lines at all. Such meagerness prompted Schmidt to publish in the *Astrophysical Journal*[107] what he now refers to as a "letter of complaint" about the lack of identifiable features. Yet, unbeknownst to him, the jackpot—a radio source named 3C 273 that was not on Matthews's high-priority list—lay just around the corner.

As we have seen, in the early 1960s positions of many discrete radio sources were often known to only about 10 arcminutes, an accuracy that corresponds to a third the apparent size of the full moon. Finding optical counterparts to these sources—especially in areas of the sky crowded with stars and faint galaxies—was sometimes impossible. Then there was a noteworthy event: three successive lunar occultations of 3C 273—the seventh-brightest radio source in the sky—provided a unique opportunity for the Australians to refine their coordinates of 3C 273 with the 210-foot Parkes radio telescope. Since astronomers know the position of the moon's rim at any time with high accuracy, the exact location, size, and structure of the occulted radio source depends only on the timing accuracy of its disappearance and reappearance.

The radio data revealed that 3C 273 was a complex radio source with two components, A and B, 20 arcseconds apart. Requesting optical follow-up observations from Palomar, Australian radio astronomer Cyril Hazard sent the new coordinates to Matthews, who conveyed them to Maarten Schmidt. Matthews also asked Sandage to take a photograph of the 3C 273 field with the 200-inch telescope. His plate revealed two optical features. The first was a jetlike filament that appeared to extend from a 13th magnitude blue starlike object, its end coinciding with Component A. The second feature—the "star" itself—coincided with Component B.[108]

Although appearing very much like a star, this object held the keys to exploring the distant universe. A recent Hubble Space Telescope image of the heart of 3C 273 clearly shows a jet of material that appears to emanate from the "star." Credit: ESA/Hubble and NASA.

As a boy in Groningen during World War II—his town in blackout so the Allied forces wouldn't drop bombs—Maarten Schmidt spent many a night with a telescope in the top room of his house observing stars as they were being occulted by the moon. In December 1962—two decades later and by then an experienced spectroscopist—he carried Hazard's fresh occultation coordinates of the peculiar source 3C 273 to the prime-focus cage of the 200-inch telescope. The cage was crowded with a spectrograph, a metal chair, night-lunch box, thermos, cables, a telephone for communicating with the night assistant, a light-tight box of unexposed photographic platesm and—most importantly—an electrically heated Air Force flight suit for warmth. Schmidt must have felt like an Air Force test pilot wedging himself into the small cockpit portrayed by Thomas Wolfe in *The Right Stuff*: "You don't get into it, you *wear* it."[109]

QUASARS: WOLVES IN SHEEP'S CLOTHING

Drawing by Russell W. Porter. Courtesy Division of Physics, Mathmatics and Astronomy, California Institute of Technology.

THE CAGE: His tall frame folded into the cramped quarters of the prime-focus cage, Maarten Schmidt felt that he was in direct contact with the universe. The cage was perched at the top of the tubelike structure that is grounded by the 200-inch mirror, and it nearly touched the apex of the dome. Seeing the sky while he observed—especially from an exotic place such as the cage—instilled in him a strong sense of mystery. He hunched on the small, hard metal seat for hours at a time as the cold penetrated into the core of his body. Schmidt mostly stared down at the crosshairs embedded in the eyepiece to guide the telescope on his object. But once in a while he would twist his head upward and gaze at the sky, partly to judge the weather and partly as a diversion from hours-long guiding. To Schmidt, it was an exquisite experience to be in the cage—he even considered it *romantic*. For one thing, it was solo work not shared with anyone. He was all by himself high above the ground, the icy cosmos breathing down his neck.

Schmidt was of the opinion that stars could not be radio sources. So he immediately dismissed the 13th magnitude blue object as an accidental foreground star. He decided to take its spectrum first "just to eliminate it from consideration." While preparing for the night of observing, he had concluded that the nearby wisp was "confoundedly faint." It would, therefore, be enormously difficult to observe unless he masked the scattered light from the nearby star.

Schmidt believed in pushing the telescope to its limits and in doing work on very faint objects, which required the world's largest mirror. A typical observing night would consist of two hours taking a spectrum of a relatively bright object, and 8 to 10 hours taking one of a faint object. Around midnight, Schmidt would stop the exposure, ride an elevator down to the observing floor, and the night assistant would observe for around 40 minutes to give Schmidt a respite. Since he stopped observing in the prime-focus cage in 1969, he has keenly missed it.

Accustomed to working on galaxies that are dozens of times fainter, Schmidt misjudged the exposure time for the blue "star" and badly overexposed its spectrum on the photographic plate. Still, his attention was caught by the cutoff of ultraviolet light in the spectrum. Was it perhaps sharper than normal? If so, did it mean something? Not comprehending the full potential of that spectrum, he waited two nights to retake it. A second, much shorter and lighter exposure revealed five broad emission lines superposed on a rather blue continuum. Their wavelengths did not correspond to any emission lines normally seen in stars, so Schmidt asked his colleague J. Beverly (Bev) Oke to observe the object at longer wavelengths. Oke had an upcoming observing run on the Mount Wilson 100-inch telescope, and his photoelectric scanner there could record photons in the near infrared at wavelengths inaccessible to photographic plates.

Not only did Oke's photoelectric scans over the wavelength range 3,300–8,400 angstroms confirm the emission lines and blue continuum recorded by Schmidt, they also yielded *another* strong emission line in the near-infrared at 7,590 Å. The relatively flat observed spectrum bore no resemblance to a black body, nor to any normal star. Oke claimed that "at least some of the optical continuum radiation must be synchrotron radiation," the only physical process understood at that time to produce that energy distribution.[110] In all, several hours spent observing at two major California telescopes had produced five emission lines on Schmidt's blue-region spectrum and one additional line in Oke's near-infrared scan. Six historic emission lines.

That is where things stood until February 1963, when the Australian radio astronomers decided to publish their occultation results for 3C 273 in the journal *Nature*. They urged Schmidt to join them in publishing, even though he didn't yet understand the nature of the emission lines in the starlike object.[111] Under pressure to write the article,

Schmidt reexamined the spectrum. Fresh at the task, this time he thought he noticed a pattern in three of his spectral lines plus the one found by Oke: going from the red to the blue, the lines appeared to get closer together and fainter.

Intrigued, Schmidt computed wavelength ratios with his circular slide rule, but unwittingly made an error in one of his calculations. Taking a second route, he decided to check the apparent regularity of the lines by comparing their wavelengths to those of the nearest Balmer lines of hydrogen, which is, after all, the most abundant element in the universe. The wavelength ratio of Oke's line at 7590 Å to the reddest laboratory Balmer line at 6562.8 Å—known as Hα—was 1.158. Each of his three lines, when compared to the next Balmer lines, also had the ratio 1.158. By then, Schmidt—in a now-celebrated moment—realized that he was seeing emission lines of glowing hydrogen gas that had been highly redshifted. Suddenly, it became clear that the simplest explanation was that the "blue star" was moving away from Earth at nearly 16% the speed of light!

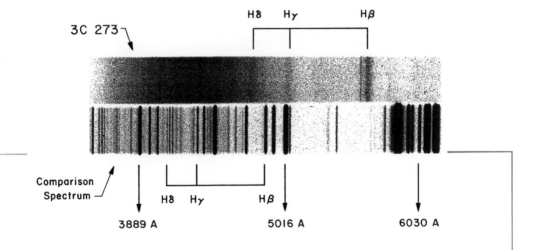

The optical spectrum of the 13th magnitude blue stellar object that was later identified as a quasar named 3C 273. The broad hydrogen emission lines in the quasar spectrum (upper strip, appearing as vertical black lines on this plate negative) are shifted to redder wavelengths compared to hydrogen lines for a source at rest in the comparison lamp spectrum (lower strip). The broader widths of the quasar lines are due to fast-moving gas. Maarten Schmidt took the spectrum on a tiny half-inch-long photographic glass plate using the prime-focus spectrograph of the 200-inch telescope on December 29, 1962. Photograph courtesy of Maarten Schmidt.

Chapter 6

Schmidt was stunned by his discovery, and not just because of the size of the measured redshift. After all, Minkowski had recently announced that the radio source 3C 295 had a redshift of 46% the speed of light, and nobody doubted it was due to the expansion of the universe. What stunned Schmidt was that, whereas Minkowski's object was very faint—20th to 21st magnitude—and obviously a galaxy, the blue object associated with 3C 273 was nearly 1,000 times brighter—13th magnitude—and yet resembled a star. If its redshift was cosmological, it meant that this starlike object was moving away from Earth at 47,000 km/s, a hundred times beyond the speed limit for binding stars to the Milky Way's gravitational field. It also meant that at more than 2 billion light-years from Earth, 3C 273's dispersed light was still capable of scorching Schmidt's photographic plate. The blue star was as luminous as one hundred normal galaxies!

A bit shaken, Schmidt tested his ideas on colleague Jesse Greenstein, who had recently submitted an article on the optical counterpart to the radio source 3C 48. Greenstein fetched a copy of his wavelength measurements and the two astronomers calculated that 3C 48 was whizzing away from Earth at 37% the speed of light[112] and was 3.5 billion light-years away. Greenstein immediately retracted the journal article in which he had mistakenly disavowed the interpretation of a high redshift to explain 3C 48's optical spectrum.

On that first afternoon, Schmidt, Greenstein, and Bev Oke, who had by then joined the loud, animated discussion, spent several hours at Schmidt's blackboard searching for a way out of the high-redshift interpretation. They attempted, through trial and error, to fit the lines of 3C 273 with wavelengths from other plausible ionized atoms, and failed. They pondered whether the redshifts might be of internal gravitational origin, instead of cosmological, but Schmidt used spectroscopic arguments to show that gravitational redshifts could be excluded.

The three astronomers rifled through the available body of astrophysical knowledge and theories to consider the observed flat shape of the radio continuum, the inferred synchrotron emission far into the ultraviolet, the computed strength of the magnetic field, and the implications of the rapid light variability that Sandage had documented in 3C 48. Problems new to astrophysics arose from the inferred high gas density and enormous energy requirements. Yet the conclusion that both optical point sources at the locations of 3C 273 and 3C 48 must have extraordinarily high redshifts stood firm. The cosmological hypothesis won out, and by 1963—only 14 years after its commissioning—the Hale telescope had reached and even exceeded the most optimistic prediction of its potential thrust into the distant universe.

Agony and Ecstasy

With the deadline looming for the March 16, 1963, issue of *Nature* magazine, it was decided that Schmidt would announce the redshift of 3C 273; Oke, the crucial near-infrared line in 3C 273; and Greenstein and Matthews would write up 3C 48. Sandage, who with Matthews had been part of the discovery of 3C 48, and had early on shared his 3C 48 data—the direct photographic plate, the photoelectric photometry, and the spectrogram—with Greenstein, and his 3C 273 plate with Schmidt, was left out of the loop. As chairman of the astronomy department, Greenstein presumably rationalized his decision as wanting to put Caltech astronomy on the map with this revolutionary announcement, without sharing credit with any other institution or astronomer.[113]

Maarten Schmidt titled his article "3C 273: A Star-like Object with Large Red-shift." It was straightforward, brief, and betrayed high confidence in the spectroscopic data. To soften the blow of his radical conclusions, Schmidt wrote, "the explanation in terms of an extragalactic origin seems most direct and least objectionable." He confided that the week following submission was a very tense and sleepless one. He thought hard about whether he had made any unwarranted assumptions or "dumb" mistakes. Had he dug himself into a hole? If so, people might think that he had been "crazy enough to be at the summit of Palomar and yet to believe that a 13th magnitude *star* was at 2 billion light-years distance—and moving away from Earth at 16% the speed of light!"[114] Yet there was nothing he could do about it: the data were convincing. Alternatively, when not worrying, Schmidt—a bred-in-the-bone explorer—felt exhilarated by the thought that amazing things at even higher redshifts must lie in wait.[115]

Sandage received news of Schmidt's breakthrough only after his own paper with Matthews on the "Optical Identification of 3C 48, 3C 196, and 3C 286 with Stellar Objects"[116] had been submitted to the *Astrophysical Journal*. Sandage did not want to be derailed from publishing hard-won data on the photometry and peculiar energy distributions of these objects. Holding on to the premise that 3C 48 was likely a nearby radio star within the Milky Way galaxy, he and Matthews appended some paragraphs that addressed—but did not embrace—the new evidence. They wrote that, if Schmidt's identification of the redshift were to be correct, the energy requirement for 3C 48 was "becoming uncomfortably large" and exceeded that of any known galaxy. Furthermore, because the brightness of 3C 48 varied, the size of the emission region had to be smaller than the distance light travels during the period of variation. Otherwise, the variations would overlap and be smeared.

Soon things became even more uncomfortable. An inspection of sky patrol plates regularly taken since the late 19th century and archived at Harvard College Observatory revealed that 3C 273 also fluctuates in brightness—over periods as short as one month.[117] The implications were shocking, yet familiar: The source was a mere one light-month in diameter, which is less than one millionth of a typical galaxy's 100,000 light-year diameter! A calculation using the inverse-square law of light propagation showed that, although from Earth 3C 273 appears relatively faint, it is among the universe's most powerful sources of radiation: it continuously belches a hundred galaxies' worth of light. What could such an object be, compact yet gaseous and trillions of times more luminous than our sun?

At first, Schmidt's announcement of his avant-garde interpretation of the object's redshift received relatively muted attention. The astronomical community needed time to digest the extraordinary consequences of this announcement. Things changed in April 1965, when Schmidt published an article in the *Astrophysical Journal* entitled "Large Redshifts of Five Quasi-Stellar Sources" to report the discovery of another radio source with a redshift of 2.01.[118] Not only astronomers but a much wider public were awed by the fact that the light that struck the 200-inch mirror had been emitted when the universe was roughly one-quarter its present age—long before Earth had formed. Suddenly stories sprouted in the popular media about how these remote objects rocked the paradigms of astronomy, physics, and philosophy. Newspapers published full-page articles about the new redshift record and how observing objects far away meant exploring the early universe. *Time* magazine began a flurry of interviews, photography, and late-night phone calls to Schmidt until his pensive portrait appeared on the cover of its March 11, 1966 issue. Inside, the article entitled "The Man on the Mountain" made Schmidt famous for his pioneering work on what were then beginning to be called quasars.[119]

A New Constituent of the Universe

Allan Sandage was frustrated that positions of radio sources were still not accurate enough to allow astronomers to locate the optical counterparts in a sky full of point sources. Quasars—with the exception of a few with wisps—and stars appear the same on single-color photographic plates. Having just discovered that the first few quasars were all unusually blue, Sandage initiated a novel search. He set the 200-inch telescope to the coordinates of each cataloged radio source and exposed a photographic plate twice, first through a blue filter and then, after a small positional offset, through an ultraviolet filter. This technique made the bluest starlike objects stand out through their high ultraviolet-to-blue ratio. To his surprise, Sandage saw several times more objects than he had expected. He was baffled that most of them did not correspond to cataloged radio positions: they *looked* like quasars, yet they had not been found during radio searches. Certain that he had discovered an entirely new class of objects—blue, starlike, and radio-*quiet*—which outnumbered quasars, Sandage wrote up his results and rushed his manuscript in time for the May 15, 1965, issue of the *Astrophysical Journal*. His scramble is recorded in the acknowledgments, which read: "It is a pleasure to thank ... Helen Czaplicki for tying [sic] the manuscript under great pressure." The journal's editor immediately published his paper without the requisite peer review.[120]

> **QUASARS:**
> These newly discovered objects were not stars, yet they looked like point sources on photographic plates. The descriptive but clumsy term "quasi-stellar radio sources" was adopted by Maarten Schmidt and Thomas Matthews in a 1964 paper. Then Allan Sandage discovered similar but radio-quiet sources and named them "quasi-stellar objects." An exasperated astrophysicist, Hong-Yee Chiu, in a review article for *Physics Today* in 1964, coined the term "quasar," to everyone's relief. The stuffy *Astrophysical Journal* accepted the new term only several years later. Often these days—including here—the two original classes are lumped together under either the general moniker "quasars" or the acronym QSOs.

The paper generated much initial excitement, but as his claim percolated through the astronomical community, Sandage learned a bitter lesson: he had neglected to account for other well-known and similarly blue and bright objects orbiting within the confines of our own Milky Way galaxy. Indeed, follow-up spectra revealed that his sample was significantly contaminated with nearby white dwarfs and blue straggler stars. In spite of this contamination, Sandage's fundamental discovery stood its ground: among verified quasars, not all are strong sources of radio waves. In fact, radio-*quiet* quasars outnumber familiar radio-*loud* quasars by about ten to one. Sandage named the new class "quasi-stellar objects," or QSOs. Since quasars selected by their strong radio emission turned out to represent only the tip of an iceberg, Sandage's color technique greatly accelerated the identification of new quasars. It provided a tool for astronomers to search directly for QSOs in abundance rather than relying on indeterminate radio positions to find them one at a time. Radio-quiet quasars would turn out to be significantly less luminous than radio sources, for reasons that would become clear only later. Although he had no idea exactly how far, how intrinsically bright, or how large these point sources were, Sandage had clearly turned up a rich new class of objects.

It is a bit humbling to realize that pointlike images of quasars—because they mimic blue stars in our Galaxy—had been photographed for nearly a century without quasars being recognized for what they are. Now, with access to a much wider band of cosmic radiation, ranging from the ultraviolet to radio wavelengths, astronomers saw a radically different view of the sky—and a hint that there may be energy sources in the universe more powerful than the nuclear furnaces at the centers of stars. Schmidt confided that it "looked like we were seeing a whole new universe almost, that you didn't know existed."[121] Objects popped out that had never before been recorded.

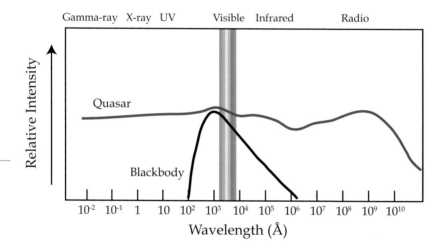

The distribution of energy in a quasar spectrum is distinct from that of a star. The quasar remains bright from radio to far ultraviolet wavelengths, whereas a star's brightness drops off on either side of a maximum determined by the star's surface temperature. Since there is no single temperature associated with their spectrum, quasars are considered "nonthermal" sources. Whereas stars tend to dominate the blue to visible and infrared wavelengths, quasars are the brightest objects in the sky at low-energy radio wavelengths and at high-energy X-ray and gamma-ray wavelengths. This distinction makes it possible to select for quasars with color surveys. Credit: Courtesy R. Schweizer.

Sandage's primary interest was stellar evolution and cosmology, with the identification of radio sources and quasars and their photometry initially a side issue. Yet the extension of the Hubble diagram to more and more distant galaxies was taking enormous amounts of his time. Soon he recognized that radio galaxies could extend his ability to plot the redshifts of distant galaxies against their magnitudes, a plot he hoped would reveal the curvature of the universe. So he continued to measure photoelectric magnitudes and colors not only of quasars but also of normal radio galaxies, until nearly all radio sources in the Third Cambridge Catalogue had been identified and measured in both redshift and luminosity.

Fuzzy Wuzzy Was a Quasar

Out of the initial attempts to identify the optical counterparts to radio sources had come quasars, a revolutionary concept that roared through the astronomical community and fueled ink pens and broadcasts in the public domain. By the late 1960s, astronomers had yet to understand the exact relationship between quasars and galaxies. Although some quasars appeared starlike, many were surrounded by wisps and fuzz so faint that astronomers trying to measure these surrounding features were frustrated. Exploring their nature was important for two reasons: linking quasars to distant galaxies would confirm that quasar redshifts were due to the cosmic expansion, and it might help astronomers understand the physical and dynamical relation between the galaxies and the quasars they may harbor.

The distances to quasars had been an open question since their nature was first discussed in the light of their large redshifts. A few diehards believed that they were local objects, possibly within our Galaxy, because their huge energies could otherwise not be explained. Although the local hypothesis had been weakened by the discovery of variability in the nuclei of some well-resolved nearby galaxies, in the early 1970s Caltech astronomer Jim Gunn provided more conclusive—and direct—evidence concerning the quasars. He located a starlike object that seemed to lie within a small cluster of galaxies and verified, with spectra taken at the 200-inch telescope, that it had the blue color and emission lines of a bona fide quasar. He then found that the redshift of this quasar was essentially the same as that of the brightest galaxy in the cluster.[122] Additional probabilistic arguments convinced him that this was not a coincidence. Hence, "at least one *real* quasar" was in the cluster, and therefore its redshift—and distance—were cosmological.

A second, more direct piece of evidence came from a meticulous observational study by Carnegie astronomer Jerome Kristian, published in 1973. He noticed that quasars seem to be extreme versions of the nuclei of nearby N galaxies and Seyfert galaxies, as judged by their spectra, colors, and variability. He decided to test this similarity by asking: "Do quasars also occur in the nuclei of galaxies?" Sandage had already calibrated how the apparent sizes of galaxies change with distance.[123] Kristian then measured the sizes of the burned-out images of 26 quasars on photographic plates taken with the 200-inch telescope. To be visible on such a plate, the image of a quasar's hypothetical host galaxy must be larger than the quasar image. He predicted that the 200-inch telescope should easily detect the fuzzy outskirts of a normal-sized galaxy underlying the relatively nearby quasars (with redshifts less than $z \approx 0.3$). For more distant quasars, the fuzz would probably remain hidden within the burned-out quasar image.

To test his hypothesis, Kristian pushed the 200-inch telescope to the limits of its resolution and photographed quasars at the Cassegrain focus. His conclusion is elegantly presented within the short abstract of his publication: "Those quasars which are predicted to show underlying galaxies do so, and those which are predicted not to show underlying galaxies do not."[124] Thus, Kristian opened the possibility that quasars lie at the centers of normal galaxies, whose fuzzy outskirts can be overwhelmed by the quasar's intense light. As an example, the 13th magnitude quasar of 3C 273 is 250 times brighter than its 19th magnitude underlying galaxy. The question then became, How ubiquitous was this relationship between quasars and galaxies?

Despite Kristian's solid evidence, his colleagues Todd Boroson and Bev Oke felt that the association between galactic fuzz and quasars was still based only on circumstantial—as opposed to spectroscopic—evidence. For example, the fuzz surrounding 3C 48 was considered to be unusually large and luminous for a normal galaxy. Yet there was no evidence of the starlight that would indicate the fuzz was a galaxy. Fortunately, technology took a great leap forward in the late 1970s as Oke and Gunn installed the "double spectrograph" on the 200-inch telescope. The spectrograph, still in use, records red and blue spectra simultaneously with a CCD detector, yielding good sky subtraction and high signal-to-noise ratios.[125] Boroson, a recently arrived Caltech postdoctoral fellow, suggested that the double spectrograph be trained on quasar 3C 48 for its maiden run, hoping that the instrument could separate the light of the quasar from its faint surrounding fuzz. The trick was to get good digital spectra of the bright starlike core of 3C 48 as well as its fuzz to the north and south, and then figure out a way to remove the light of the quasar that overflows and contaminates the light from the fuzz.

In 1982 during a night of pristine seeing when Earth's atmosphere was especially tranquil, Boroson and Oke set to work. The spectrum of 3C 48 showed that the fuzz is a galaxy with the same redshift as the quasar.[126] This left no doubt that quasars are highly luminous parts of galactic nuclei. However, the astronomers were astonished to see strong hydrogen absorption lines—signatures of young, recently formed stars—in the fuzz. It turns out they were looking at a disturbed galaxy in which a massive burst of star formation had occurred a few hundred million years earlier. Because of the highly luminous fuzz and the strong hydrogen absorption lines, the host galaxy of 3C 48 was clearly not the normal elliptical they had expected. It was not a normal galaxy at all.

The discovery of absorption lines from stars in the fuzz around 3C 48 provided a crucial link between quasars, active galactic nuclei, and disturbed galaxies. However, this link provoked yet another question: Which came first, the galaxy or the quasar?

Quasars Throughout the Universe

The discovery of quasars completely changed Schmidt's research focus. There was so much to investigate that he became fascinated by them and, he jokes, "I apparently got stuck to them."[127] Over the next few years, he phased out his work in other research areas. Galactic star formation, recycling, and chemical enrichment were replaced by his new fascination: the distribution of quasars in the universe. By 1965 the most distant quasar known was 3C 9 at a redshift of 2.01. It more than quadrupled the former redshift record of 0.46 held in 1960 by 3C 295, Minkowski's radio galaxy. Astronomers now faced the daunting prospect of exploring a volume of the universe extending to 10 billion light-years from Earth. Yet quasars were relatively rare objects. With the trickle of quasars being laboriously identified, it soon became clear that there would have to be a lot of groping blindly for quasars unless Palomar astronomers played their trump card: surveying the sky with wide-field cameras.

The Third Cambridge Catalogue of Radio Sources (Revised), published in 1959, listed around 300 radio sources over the northern sky down to a certain flux limit. Some of these had by now been identified as quasars and observed to magnitudes as faint as around 18.5. Instead of spinning his wheels with random radio sources passed along to him, Schmidt—imbued with a solid knowledge of classical astronomy—decided to select a statistically controlled subsample of these known quasars. His goal was not to attempt to understand them better physically, but to take a kind of census of their whereabouts in the universe, independent of *what* they are.

CHAPTER 6

"Sometimes when I hear an artist explain their work, I find that I am not enlightened, because it is the art itself which speaks. And, somehow, I prefer not to try and explain what I felt [about my own work] because I am not sure that I really know. I feel that whatever happened speaks for itself. After around 60 years of research, I don't ask any more why I'm motivated. It's always the same: I know there's a universe and I have to investigate it."

(M. Schmidt, personal interviews with the author, 2007–2010; photo by author)

Based on spectroscopic and photometric observations by himself and by Sandage, Schmidt statistically analyzed the distribution of 33 quasars in space. His startling conclusion was that their space density increases steeply with redshift, which means toward earlier times in the universe. And not by a small amount: there were around thirtyfold more quasars per so-called "co-moving" volume 6 billion years ago, at a redshift of 1, than there are now.[128] Schmidt had no inkling whether—or at which redshift—the steep rise in quasar density might reach a peak. He clearly needed a larger sample.

Happily, Richard Green, one of Schmidt's graduate students, decided in 1972 to conduct for his PhD thesis the first complete all-sky survey of bright quasars. Though Schmidt warned him that this was too big a project, Green sat night after night with zest at the guide scope of the Palomar 18-inch Schmidt camera (the same one that Zwicky had made famous). He photographed what eventually amounted to one quarter (or 10,714 square degrees) of the entire sky above galactic latitude 30°. His search method capitalized on the camera's large, 8.5-degree-wide field of view and on Sandage's finding that—even though the vast majority of quasars do not emit radio waves—they can all be identified by their blue colors. Green's discovery technique had been pioneered by Fritz Zwicky in his search for faint blue stars more than 25 years earlier.

Green took a total of 266 photographs, each doubly exposed through ultraviolet and blue filters, from which he identified 1,874 blue starlike objects. The small scale of the photographs did not allow him to distinguish stars from distant small galaxies, so he and Schmidt spent many nights at the 200-inch telescope taking follow-up spectra to classify the objects of interest. Their reward was the discovery of a "bunch" of new quasars every few weeks. The admirable 10-year labor uncovered not only more bright quasars—very few of which were known from radio surveys—but also other types of exotic and interesting objects such as blue stars, binaries, magnetic variables, and prototypes for new classes of objects. The Palomar-Green catalog,[129] published in 1986, still serves as a resource for astronomers. A subset of the catalog, dubbed the Palomar Bright Quasar Survey, was the largest survey of quasars at the time.

In 1983, Schmidt and Green statistically analyzed their large new sample of bright quasars, and confirmed Schmidt's earlier finding that the number density of quasars increases steeply toward earlier times in the universe. In addition, the number of high-luminosity quasars increases faster than does the number of low-luminosity quasars, an unexpected result that still stands.[130] Although it may appear counterintuitive, the highest-luminosity quasars formed earlier than the lower-luminosity quasars on average. This was an early piece of evidence that the contents of the universe are evolving, and doing so in surprising ways.

How did quasars behave in the first few billion years after the big bang? Did they continue to increase in number with increasing redshift? There were rumors of a mysterious "quasar cutoff" at high redshifts. Yet quasars were so hard to find that few were known beyond a redshift of 2. Hungry for more data, Schmidt wanted to push further into quasar territory, and deeper into space, with the 200-inch telescope. This time he was going to search not for bright quasars, as Richard Green had done, but for the most distant quasars. He yearned to reach the cosmic era of their formation.

Chapter 6

The Rise and Decline of Quasars

About this time, in the early 1980s, Schmidt, Gunn, and Don Schneider—Gunn's former student and now Schmidt's postdoc—were conducting a survey for faint quasars at the 200-inch telescope. They were working with Gunn's brainchild, a CCD imaging and spectroscopic instrument he had named PFUEI or Prime Focus Universal Extragalactic Instrument. The survey required a statistically complete sample of the properties of a large number of distant quasars, and quasars are rare. In addition, the field of the CCD chip in PFUEI was tiny, so progress was slow. As the three astronomers were dining at the Monastery one evening, Schmidt asked: "Is there any way we can use this instrument to look for really distant quasars?" And then: "I am wondering, couldn't we stop the telescope and just let the sky drift by?"[131] Gunn sat contemplating this query, pressing the heel of his hand to his forehead and rocking back and forth. After about two minutes, he answered confidently "Yes." He refigured PFUEI's electronics and software and invented the idea of the transit survey, all in time for an observing run less than two months later.

In a transit survey, astronomers don't take images of individual fields in the sky. Instead, they park the telescope and let the sky drift by overhead. This requires turning off the drive motors with which the 200-inch telescope normally tracks objects in their diurnal motion. As the oil pumps are shut down as well, the reassuring hum of the motors is replaced by a strange, eerie silence in the dome. In continuous-readout mode during hours-long observations, the CCD records long swaths of the sky as Earth rotates like a motorless tracking engine that is more accurate than any human-built motor. Gunn had inserted a grism—a prism with a diffraction grating on one surface to enable imaging and spectroscopy of the same field of view. The grism dispersed the light from stars, galaxies, and quasars into small spectra. Among the many thousands of spectra that were recorded each night on reels and reels of magnetic tape, those of quasars stood out most. Their spectra have a blue continuum with strong, broad emission lines, especially of Lyman-α. This was a very different technique from the double images of Green's and Sandage's searches.

During this time, Gunn was working on an instrument he named "4-Shooter," a ground-based prototype for the first Wide Field and Planetary Camera to be deployed on the Hubble Space Telescope. 4-Shooter featured a pyramid of four mirrors in its focal plane that split and deflected the incoming beam of light onto four CCD detectors. These covered four times more contiguous sky than did PFUEI. The success of PFUEI's transit mode inspired Gunn to design 4-Shooter as a transit instrument from the outset. As soon as it was ready, it was installed at the Cassegrain focus beneath the 200-inch mirror.

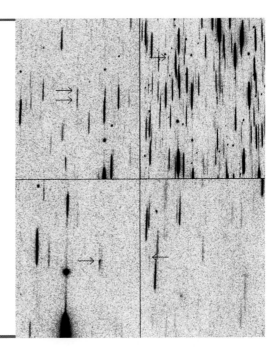

This image shows four fields from the grism transit survey with 4-Shooter. Each 4 arcminute × 5 arcminute quadrant (240 × 300 pixels) contains a high-redshift quasar among the stars and galaxies. The point sources are zero-order images and the streaks correspond to their spectra. The arrows mark the quasar emission lines—prominently Lyman-α— found by a search algorithm. The frames are precisely oriented with east at the top and north to the right. The direction of sky drift was from top to bottom, with the blue part of the spectrum at the top, red at the bottom. Figure 2 from D. P. Schneider, M. Schmidt, and J. E. Gunn, "Spectroscopic CCD Surveys for Quasars at Large Redshift. 3: The Palomar Transit GRISM Survey Catalog," *Astronomical Journal* 107 (1994): 1245–1269. © AAS. Reproduced with permission.

On a photographic plate, the transit mode would have produced an indecipherable jumble: every object would have left a streak as it passed over the telescope. But since CCD detectors can be read out electronically at the same rate at which the image drifts by, the software was able to compensate for the object's motion across the chip. This clever technique had been invented by Air Force scientists for various purposes of their own, such as imaging the ground from moving spy satellites or aircraft as the ground sped by under the camera. Fortunately, it worked just as well astronomically.

The goal of Schmidt, Gunn, and Schneider was not just to find the highest-redshift quasars, but to characterize their optical properties, space density, and radio loudness. Collaborating for over a dozen years, they found—through sustained efforts— hundreds of distant quasars with redshifts between 2.7 and 4.9. In 1995, they published a detailed statistical analysis of their data.[132] Instead of the rumored cutoff, they found solid evidence that the space density of quasars peaks around redshift 2.5 and then gently declines toward higher redshifts. But the space density does not vanish. Its peak at redshift 2.5, when the universe was only one fifth of its present age, signals the epoch at which most quasars were active, probably driven—as we now understand—by frequent collisions and mergers of gas-rich galaxies in the dense, young universe.

Yet that is not the end of the story. Understanding the details of the rise, peak, and decline of quasars and of their relationship to galaxies was a major challenge for theories of galaxy formation and evolution. Why did the most luminous quasars appear before the less luminous ones? And what was the relation between the formation of quasars and the formation of galaxies?

Answers to these questions would begin rolling in shortly after that major 1995 paper was published. By then the Hubble Space Telescope's initially marred optics had been corrected, and extremely deep images of the distant universe were obtained with it. Also, a new generation of ground-based 8- to 10-meter telescopes could record spectra of objects past redshift 6. Much work since then has not only confirmed the pioneering results by Maarten Schmidt and collaborators, but has also begun yielding broader and deeper insights into the mysterious engines that drive quasar activity: disks of accreting gas swirl around supermassive black holes, heat up, and eject enormously powerful jets of relativistic plasma through electromagnetic forces.

Water droplets on the surface of 200-inch mirror during cleaning. Photo by the author.

7

Piercing the Galactic "Smog"

Chapter 7

Absorbing material, such as sinuous dark clouds in the center of our Milky Way galaxy, shown here photographed with the Hubble Space Telescope, are a hindrance to the exploration of the universe. Astronomers observing at visual wavelengths cannot see what lies behind 30 magnitudes of extinction: they receive only one trillionth of the light emitted from sources located behind the clouds. However, in the late 1950s and early 1960s, there was a unique confluence at Palomar of two new observing techniques. The radio astronomy revolution was just a decade old when a group of physicists began to pry open a second window to exploration: the infrared. The beauty of new wavelength portals is that they reveal physical processes with a fresh perspective. The infrared portal, which bridges the electromagnetic spectrum between optical and radio wavelengths, offers at least three advantages over the optical portal. First, it reveals cold states of matter—the molecules and solids found in low-temperature environments such as planetary surfaces, star-forming molecular gas clouds, and some stellar atmospheres. Second, infrared light pierces dust clouds that obscure stellar nurseries, aging stars, and active quasars, rendering them visible. Third, the cosmic expansion stretches the ultraviolet light emitted by distant galaxies to longer wavelengths accessible with ground-based telescopes, so infrared astronomers can study newborn galaxies and the early universe. Credit: ESO/F. Char.

While some Palomar astronomers were hunting down quasars 10 billion light-years away, others were frustrated by their inability to view the center of our own Galaxy a mere 26,000 light-years away—half a million times closer than the quasars. Although certain quasars shine a thousand times brighter in visible light than does the entire Milky Way galaxy, that is not the whole story. The disk of our Galaxy is heavily polluted with light-absorbing dust. The sun and our Earth lie in this disk and are, therefore, surrounded by a cosmic "smog." Optical photons from our Galaxy's bright nucleus seeking a way through the disk are attenuated to near-invisibility: less than one in a trillion optical photons arrives at Earth from the galactic center. The remainder are detoured by scattering or devoured by absorption in interstellar dust clouds. Photons at longer wavelengths fare better in piercing the dust: at an infrared wavelength of 2.2 microns, about one in ten photons arrive at Earth from the galactic center. Pointing a telescope *perpendicular* to the galactic disk provides a relatively clear view of more distant objects. But nature conspired against us there, too, since distant cosmic light reaches us at wavelengths longer than could be detected by existing instruments. That was the situation in the early 1960s, but things were about to change.

Far Sighted

In 1963, Caltech physicists Robert Leighton and Gerry Neugebauer were intrigued by the promise of the few objects observed to date at infrared wavelengths. They wanted to find out what the northern sky viewed at 2.2 microns would look like. The enterprising physicists crafted a 62-inch-diameter parabolic-shaped dish by rotating a vat full of epoxy until it solidified. They outfitted their newly manufactured telescope with an array of eight lead sulfide detectors sensitive to near-infrared light, then hauled the construct—along with various graduate students—to Mount Wilson Observatory. They scanned a large chunk of the northern sky at 2.2 microns "just to see what is out there"[133]—a trademark Leighton and Neugebauer rationale. The two physicists and their students and technicians succeeded beyond expectations. Although skeptical optical astronomers had predicted the survey would find only known stars, their catalog—published in 1969 and known as the "Two-Micron Survey"—listed more than 5,000 sources of which around 100 did not appear in any optical catalogs.[134] It was a unique resource for infrared astronomers for more than three decades.

Numerous hurdles had to be overcome before routine infrared observations became feasible. Technologically, detector sensitivities and filter transmissions had to be matched with windows of atmospheric transparency, and warm telescopes had to be baffled to diminish their own infrared emission. Administratively, valuable telescope time had to be diverted from somewhat resentful optical astronomers to make it all happen. In retrospect, some old hands of infrared astronomy find it amazing that it ever came about.

Except near the poles and at high altitude, everything on Earth is relatively warm, including the air we breathe and the telescopes that astronomers use. Room temperature corresponds to about 300 K, where materials and human bodies radiate a spectrum peaking at about 10 microns. Astronomical observations from Earth's surface are most difficult from 30 to 300 microns, where the atmosphere does not transmit any radiation. Bodies at 300 to 3,000 K (half the temperature of the sun) emit most of their radiation from 1 to 10 microns. The general glow from the dome, the trees outside, and even the observer and night assistant drown out faint emissions from celestial sources. The situation is therefore akin to looking at the night sky with an infrared detector placed inside a furnace.

Things are a bit easier around 1–2.5 microns, where Neugebauer and Leighton began their work. But even here, unwanted emissions from the sky, telescopes, and dome were interfering. Techniques such as beam switching were invented to subtract unwanted foreground emissions from the faint infrared signals emitted by planets, stars, and galaxies. Pioneering infrared astronomers learned how to suppress stray radiation by adding baffles at strategic locations along the light path inside the 200-inch telescope. Instruments were cooled with liquid nitrogen and the optics were rearranged so that infrared detectors would see only cold surfaces and the sky. In the end, not even a sliver of the 200-inch telescope—other than its mirrors—was visible to the detectors when imaging a patch of sky.

Even so, the calibration of modern infrared detectors and the processing of the images they produce required ingenuity. Often the astronomical objects to be studied contribute only a fraction of a percent to the total radiation registered, and the high background radiation must be subtracted with great care. When pushing to the faintest limits, ground-based infrared astronomers observe objects that may be a million times fainter than the atmospheric emission ahead of them. This is as difficult as trying to observe faint galaxies on a sunny day with an optical telescope! Despite these difficulties, the cumulative efforts of infrared astronomers would eventually expose a hidden universe and provide an enduring thrust to astronomical research.

Protostars: Galactic Vacuum Cleaners

During the winter of 1966, some theoretical astrophysicists predicted how a star in the process of forming through gravitational collapse of a gas cloud might appear. By coincidence, Caltech graduate student Eric Becklin was searching for such "protostars." He had heard about these hypothetical objects from Leighton, one of his physics professors, who told him that the Orion Nebula—known to host many young stars—was a promising place to search. The job called for a small-beamed photometer to block strong background emission from the nebula that would otherwise overwhelm any faint protostars. Becklin borrowed the infrared photometer used to follow up sources in the 2.2-micron survey and took it to the Mount Wilson 60-inch telescope.

If the Orion Nebula is viewed in visible light, only some hot bright stars such as those in the Trapezium star cluster and a few dust clouds illuminated by starlight will show up in the telescope. However, with his infrared detector Becklin picked up the glow of Orion's great nebula and a bounty of otherwise hidden sources. Searching for protostars, he moved the telescope north of the Trapezium cluster toward a promising dark, dust-obscured region. Scanning there back and forth and up and down, he suddenly came upon a whopping signal as bright as those of the Trapezium stars. Although the newly found infrared source would turn out to be the brightest in the sky besides the planets, when Becklin looked through the eyepiece directly he saw no detectible optical counterpart. There was no visible star at the position.

Back in Pasadena after his run, Becklin showed his infrared scans of the bright source to astronomer Guido Münch, whose reaction was unforgettably vivid: "Dios mio, Eric! Eso es lo más importante que he visto!" ("Jesus *Christ*, Eric! That is the most important thing I have ever seen!")[135] Immediately, Münch arranged for observing time on the 200-inch telescope to figure out what was going on. He and Becklin mounted the same photometer inside the hollow east arm of the telescope's horseshoe and then went after broader wavelength coverage of the object, at 1.65, 2.2, and 3.5 microns. Colleague Jim Westphal also provided scans from the east arm at 10 microns. These critical observations proved beyond any doubt that Becklin's object was a spectacularly bright infrared point source one arcminute northwest of the Trapezium. But what was it?

Münch, an expert in star formation, helped Becklin analyze the scans. The object—about the size of the solar system—was estimated to have a temperature of 700 K, twice room temperature on Earth. Furthermore, its infrared luminosity was higher than the optical luminosity of the sun by a factor of 500, and it appeared to be embedded in the nebula rather than lying in front of or behind it. Could it be a small protostar? Or a heavily reddened young star? Münch pushed Becklin to get his arguments down firmly since others were beginning to claim that the object was a background star. In a 1967 paper, Becklin and graduate adviser Neugebauer wrote convincingly that the curious infrared object, soon dubbed "Becklin's Star," or the "Becklin-Neugebauer Object," appeared to be a massive protostar still enshrouded in its cool dusty shell.[136]

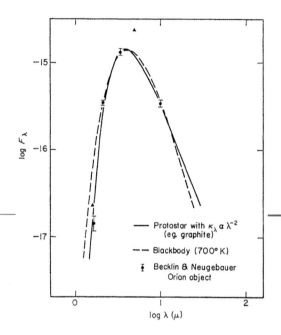

While Becklin was searching for protostars, Richard Larson—Münch's own thesis student—was investigating the dynamics of how gas clouds collapse to protostars. According to his models, in the early stages of forming—and before nuclear reactions begin—a protostar's core is heated by infalling gas and dust grains from the surrounding shell. Simultaneously, the shell absorbs energy from the core, heats up, and radiates in the far infrared. Thus, although the forming star is itself completely obscured, its cocoon of dust glows brightly in the infrared as a harbinger of its formation. Larson modeled the energy emitted by a collapsing protostar at different wavelengths (the solid curve in this figure) and compared it to Becklin's and Neugebauer's measurements of the Becklin-Neugebauer Object (the filled circles in the figure), and with the energy distribution of a black body at 700 K. Even with the idealized nature of the model and the observational errors, his plot of the data and two curves shows a surprising similarity. The Becklin-Neugebauer Object was, indeed, a protostar in the early stages of forming. Figure 7 from R. B. Larson, "The Emitted Spectrum of a Proto-star," *Monthly Notices of the Royal Astronomical Society* 145 (1969): 297, by permission of Richard Larson.

The likely culprit triggering the collapse of nearby dust and gas clouds to form objects such as the Becklin-Neugebauer Object is radiation pressure from the Trapezium stars. Since those early days, observational and theoretical evidence has accumulated that dust shrouds, strong winds (including supersonic flow of ionized hydrogen), extensive shock fronts, OH and water masers, and clouds of cool molecular gas may contribute to the formation of protostars.

Evolved Stars: Galactic Smokestacks

While following up on sources cataloged during the 2.2-micron sky survey, Becklin—still a graduate student—discovered another intriguing source: IRC+10216. Brighter at 5 microns than anything outside the solar system, the object so excited Becklin that he began to hyperventilate as he described it to Neugebauer by phone. In contrast to the Becklin-Neugebauer Object, this object lay above the plane of our Galaxy in a region devoid of gas, dust, and young stars. It also appeared slightly elongated and fuzzy on a red-sensitive plate taken by Chip Arp with the 200-inch telescope.[137] Becklin found no evidence for transverse motion across the sky when he compared the object's position on a plate taken by Arp in 1969 with a 1954 Palomar Sky Survey plate. This told him that the object was at least 100 parsecs away. But was it galactic or extragalactic?

By chance, IRC+10216 was to be occulted twice by the moon during the winter of 1970–1971. Becklin and Neugebauer jumped at the opportunity to measure the object's changing brightness as it emerged from behind the moon. Helped by some geometry, such measurements hinted at the object's physical size and configuration. Since IRC+10216's brightness decreased slowly and in steps, the object likely consisted of a small, dense, and warm inner shell surrounded by a less dense, cool outer shell.[138] Thus, the two astronomers concluded that IRC+10216 was probably a carbon star—a red giant in the late stages of evolution blowing off its sooty carbon-compound atmosphere as it becomes a white dwarf.

Follow-up observations led to the understanding that, even though the Becklin-Neugebauer Object and IRC+10216 appear similar, they represent two very different types of dust shells. Both are infrared-bright objects with featureless spectra, and both are embedded in dust heated to around 700 K. However, they do not share a similar genesis. The Becklin-Neugebauer Object, a protostar, was created from a collapsing cloud of gas and dust, while IRC+10216 is a star that created its own dust cloud as it evolved. In the future, the appearance of these two objects will diverge. The protostar will blow off its surrounding dust shell and emerge shining brightly at visible wavelengths. The aging central star of IRC+10216, on the other hand, will continue losing mass as it spews out shells of gas enriched with the products of nuclear burning, such as carbon, nitrogen, iron, and silicon. As the gas cools, these chemical elements will condense into opaque dust and provide fuel for new generations of stars. Spreading their "smog," late-type stars such as IRC+10216 are thought to be major polluters of the interstellar medium. In fact, about half of the "smog" in present-day galaxies is thought to have been contributed by such aging stars, while the other half comes from massive young stars blowing winds and from exploding supernovae.

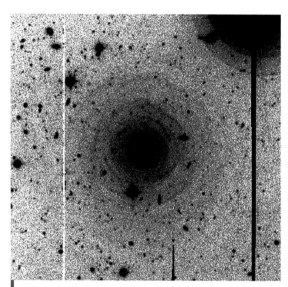 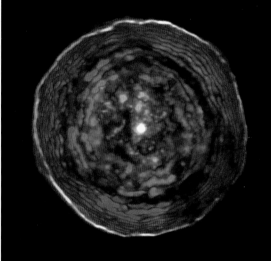

Gas tends to be recycled and enriched, as exemplified in these two examples of the shells of enriched gas being exhaled by two carbon stars. Left: In IRC+10216, several shells mark episodes of gas ejection occurring about 300 years apart. Some shells appear segmented, suggesting that the sources of outflows occupy different regions on the star's surface at different epochs. The field of view is 3.2 × 3.2 square arcminutes. Right: The carbon star U Antliae shows a shell of material expanding at high speed having have been ejected some 2,700 years ago. This image was produced from a three-dimensional dataset at multiple wavelengths. Credit, left: I. C. Leão et al., "The Circumstellar Envelope of IRC+10216 from Milli-arcsecond to Arcmin Scales," *Astronomy and Astrophysics* 455 (2006): 187–194, reproduced with permission, © ESO and Izan Leão; right: Atacama Large Millimeter Array [ALMA] (ESO/NAOJ/NRAO), F. Kerschbaum.

Galactic Center, Where Art Thou?

While the 1963 discovery of quasars generated intense curiosity about the broad range of physical processes occurring in the central regions of some other galaxies, optical observations of the center of our own Galaxy were hopeless. Dust chokes off all but one in every trillion optical photons emitted from that center. At longer wavelengths, however, the Galaxy becomes more and more transparent. Radio astronomers had mapped the distribution of its neutral hydrogen gas. As they defined the galactic plane, they discovered a strong nonthermal radio source named Sagittarius A near the dynamical center of our Galaxy.[139] Knowing the precise location of the Milky Way's center was crucial in order to pin down its size, stellar density, morphological type, and mass. These fundamental parameters, along with the ages and chemical compositions of the stars populating the central region, could be compared to those of other galaxies to model how they formed and evolved.

Neugebauer and Leighton realized just how much the clarity of view and apparent brightness of the galactic center were helped by going to infrared wavelengths, and they knew the potential value of pinpointing its exact location. Yet, due to an unfortunate confusion in the coordinates, they failed to see it on their first 2.2-micron scans in 1964 and 1965. Again graduate student Becklin developed a clever plan. First, he took the group's infrared photometer to Mount Wilson's 60-inch telescope and scanned the nucleus of M31, our neighbor galaxy in Andromeda, to measure its brightness at 2.2 microns. He then assumed the Milky Way's center to be of similar intrinsic brightness. After accounting for the estimated extinction between Earth and the galactic center, he convinced himself that the galactic center should just be visible at 2.2 microns. Though he was discouraged from trying by Neugebauer, in August 1966 Becklin defiantly returned to Mount Wilson and scanned the region near Sagittarius A. In a moment of victory, as the telescope approached the position of Sagittarius A, a bright peak abruptly appeared. Realizing that he was the first to see the galactic center in the infrared, Becklin described it as "the most interesting region, the heart of where all the action was."[140]

That impressive discovery earned Becklin and Neugebauer coveted time on the 200-inch telescope, with its higher angular resolution, better tracking, and data with improved signal-to-noise ratios. The two hunched over the output from the photometer, watching as it scanned back and forth across Sagittarius A, recording the data on a large number of strip charts. Though rudimentary, the strip chart tracings showed that Sagittarius A coincided with a bright compact region of the galactic center, presumably its nucleus. Observing the nucleus of M31 to better understand the obscured nucleus of our own Galaxy was the best that could be done at the time. Allan Sandage, Becklin, and Neugebauer saw similarly shaped light profiles in their scans of the Andromeda galaxy and the Milky Way, noting that the central region of M31 is a factor of 2.4 fainter than that of the Milky Way.[141] Sandage, upon seeing the scans, remarked that "Walter Baade would have given his eye teeth to make those measurements."[142]

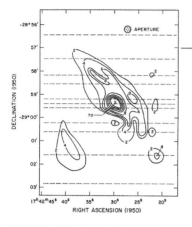

Over four decades, our view of the galactic center has been chiseled into finer detail, revealing a wide range of tantalizing physical structures and phenomena. Top: The contour map at 2.2 microns shows the first detection by Becklin and Neugebauer in 1966 of structure in the central 1 arcminute of the galactic center. The stellar density appears to increase toward Sagittarius A, the central source marked with an "x," which coincides with a strong radio source at the dynamical center of our Galaxy. The contours show how the intensity varies along right ascension (dashed lines) and declination, with a scanning aperture of 15 arcseconds. Middle: The scans were repeated 35 years later at the same wavelength and over the same region, but with a large gain in resolution, a modern 256 × 256 array detector. Bottom: A panorama (approximately 440 parsecs wide) surrounding the galactic center scanned at a wavelength of 8 microns. Three types of dust clouds are visible: bright reflection nebulae illuminated by young massive stars; dark, opaque dust globules in silhouette, and warm dust heated by radiation. Top: Figure 6 from E. E. Becklin and G. Neugebauer, "Infrared Observations of the Galactic Center," *Astrophysical Journal* 151 (1968): 145. © AAS. Reproduced with permission; middle: Courtesy of 2MASS/UMass/IPAC-Caltech/NASA/NSF; bottom: NASA/JPL-Caltech.

For the next several years, Becklin and Neugebauer continued to scan the central region around Sagittarius A at 2.2 and 10.1 microns with increasing resolution. Their 1975 paper[143] announced more complex and unexpected structures in the core. Current observations at radio, X-ray, and infrared wavelengths suggest that the core of the Galaxy contains a supermassive black hole of about 4 million solar masses, surrounded by thousands of stars moving at speeds of up to 10,000 km/s. Stars are so crowded at the centers of other galaxies that the Hubble Space Telescope cannot distinguish them.

The focus at Palomar was to identify the mechanism behind the intense infrared radiation streaming from the nuclei of certain galaxies. Was it thermal radiation from stars and warm dust? Or was it, as many astronomers supported, from a nonthermal source such as a violent event that accelerated charged particles moving in a magnetic field to relativistic speeds? If so, such particles would emit synchrotron radiation. Neugebauer, Becklin, and instrument specialist Keith Matthews hypothesized that the size and light profile of a galaxy's nucleus hold the key. That is, a point source would favor nonthermal radiation while an extended source would favor thermal radiation. However, they proceeded with the caveat that nature sometimes contrives to align thermal components, such as warm dust and cool stars, so that their combined spectrum resembles synchrotron or other nonthermal radiation.

NGC 1068, a nearby galaxy with an extremely bright nucleus, seemed a good test object for their hypothesis. Neugebauer built a single-aperture photometer for the 200-inch telescope to observe its nucleus at 10 microns. Since ultraviolet and visible light from stars are negligible at that wavelength, the light was expected to be dominated by the galaxy's core. Their scan revealed a spatially extended source out of which flowed radiation nearly three orders of magnitude greater than would be expected from stars alone.[144] Thus, they concluded that the radiation was thermal and likely produced by warm dust.

These observations provided strong evidence that dust has a significant impact on the physical processes occurring in the centers of galaxies. To Neugebauer, this handful of serendipitous discoveries—the Becklin-Neugebauer Object, IRC+10216, the nature of the galactic center, and then that of NGC 1068—illustrated the need for an unbiased survey of infrared sources. Instead of just pointing the telescope at individual promising objects, such a survey would provide a complete inventory of the infrared sky. These discoveries were cited as scientific justification for NASA's investment in a new space mission to map the sky at infrared wavelengths.

Chapter 7

From Oddballs to LIRGs

During the infrared space mission's planning stages, careful attention was paid to the lessons learned from the 1960s investigation of radio sources. At that time, the positional accuracy required to identify optically unknown sources had been difficult to achieve. Including high accuracy up front in the experimental design of the satellite would smooth the path to a rapid postmission turnaround from infrared sources to scientific discoveries. To achieve these exacting goals, Caltech engineers, physicists, and astronomers would trek every morning to the nearby Jet Propulsion Laboratory with their marching orders, deliverables, and timelines—all mission-oriented stuff. This is what it took to conduct a mission in the precise NASA way of doing business.

The resulting Infrared Astronomy Satellite (IRAS), a joint venture of the United States, the Netherlands, and the United Kingdom, was launched in 1983. While orbiting Earth over a period of 10 months, it scanned more than 96% of the sky at wavelengths between 10 and 100 microns, making it sensitive to bodies or dust with temperatures from about 300 K to 30 K. By the time observations ceased, the database of known *extragalactic* infrared sources had grown from a few dozen to 30,000, most of which were too faint at visual wavelengths to have been included in previous optical catalogs. Therefore, the first task was to match—whenever possible—the positions of these newfound sources with optical counterparts in the POSS data, and then to follow up with the 200-inch telescope to measure redshifts and characterize the sources.

In 1978, when the point-at-anything-that-looks-interesting mode was still in vogue but beginning to subside, Tom Soifer, an infrared astronomer, joined the so-named "Infrared Army" group at Caltech. The prospect of using the world's largest telescope to study objects discovered with the new NASA satellite had lured him. Soifer learned of the important statistical methods developed by Maarten Schmidt to derive the luminosity function of quasars in the 1960s: well-defined flux-limited samples and correcting biases. The boundary conditions of Soifer's and Schmidt's work were quite similar. Schmidt had worked with quasars—a new class of objects discovered at radio wavelengths—and had applied optical techniques to derive robust measures of their physical conditions.

In 1984 and flush with IRAS data, Soifer and Dave Sanders—a newly hired Caltech postdoctoral research fellow—set out to follow Schmidt's path. Their goal was to analyze the characteristics of infrared-bright galaxies, especially their emission mechanisms. Before obtaining optical spectra, the first step was to cull a sample of sources from the IRAS catalog that were (i) bright at 60 microns, (ii) neither stars nor solar system objects, (iii) paired with an unambiguous optical counterpart on Palomar 48-inch Schmidt plates, (iv) extragalactic, as confirmed by either their membership in a catalog or by a galaxy-like spectrum obtained at Palomar, (v) located well above or below the dusty galactic plane, and (vi) sufficiently numerous to calculate their space densities.

The project, carried out at Palomar with ample observing time, was a significant improvement over the piecemeal samples appearing in the literature. Soifer, Sanders, and their colleagues spent countless nights at the 200-inch and 60-inch telescopes to measure redshifts and acquire supplemental images for their sample of 324 extragalactic infrared sources.[145] Because these sources were sufficiently bright and nearby, classical optical techniques could be applied to measure the objects' distances and total infrared luminosities. The main objective was to understand the physical processes responsible for the infrared emission. At this time they had not yet realized that the *morphology* of the sources would play a pivotal role.

Soifer and Sanders worried that, other than perhaps some exotic objects, the sample would not reveal anything shocking or even new—that it might go down in history as just another statistically complete sample. So the data were carefully scrutinized for trends and signs of abnormality. One of their first goals was to find out whether the objects were as rare as luminous radio galaxies, starburst galaxies, and quasars. But which metric would make sense? The newly discovered objects emitted most of their energy in the infrared, while quasars emit most of their energy in the optical and ultraviolet. These are widely different portions of the electromagnetic spectrum. How could two objects be compared at visible wavelengths if one of them was invisible?

While on sabbatical at Cornell University to write a paper describing his team's work, Soifer learned of an existing metric that had been used more often for stars than for galaxies. The metric, known as "bolometric luminosity," is the *total* energy emitted by a source at all wavelengths. It is common to every kind of object—whether a star, a quasar, a galaxy of normal stars, or a galaxy shrouded in dust. The value of this metric now seemed obvious to him. For infrared-luminous galaxies, this meant that the far-infrared luminosity was itself a good approximation of the bolometric luminosity, since the contribution of normal stars was minor. For normal galaxies, adding up the optical and near-infrared luminosities would suffice. With this "homework" done, Soifer plotted the bolometric luminosities for his sample of infrared-bright galaxies following Schmidt's techniques for quasars.

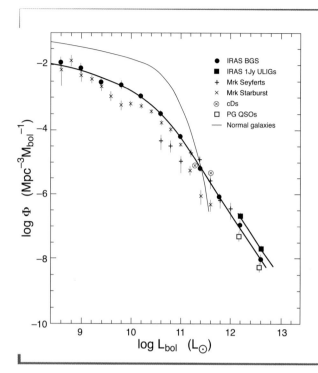

The plot shown here, conceptualized in 1986, altered our way of assessing galaxy populations in the universe. It shows the space densities (vertical axis) of various classes of extragalactic objects plotted against the objects' bolometric luminosities (horizontal axis). What makes this plot of so-called "bolometric-luminosity functions" compelling is the revelation that some infrared galaxies are more numerous than quasars in the local universe. They are also as luminous as the quasars selected during the Palomar-Green survey. Figure 1 from D. B. Sanders and I. F. Mirabel, "Luminous Infrared Galaxies," *Annual Review of Astronomy and Astrophysics* 34 (1996): 749. Reproduced with permission from D. B. Sanders, University of Hawaii.

The location of the infrared-luminous galaxies in the plot stunned the team: the galaxies with the highest bolometric luminosities emitted more than 10^{12} solar luminosities, well into the range of quasar luminosities.[146] And there was no evidence for any cutoff at that luminosity. An important consequence was that the existing criteria used to identify and group galaxies into various classes would need to be revised, since hitherto they had been based solely on optical observations. Luminous infrared galaxies, dubbed LIRGs, turned out to be not just a few oddballs, but a whole new class of objects that rival quasars in bolometric luminosity.

This discovery demonstrated the necessity of using a common yardstick to measure and compare different types of galaxies. Maarten Schmidt had defined a quasar as having a minimum visual luminosity. Soifer and Sanders consulted with Schmidt, telling him that their most luminous galaxies were equal in bolometric luminosity to his quasars. Not only that, in the same volume of space, there were three times more bright infrared galaxies than optically selected quasars from the Palomar-Green survey.[147] In fact, today we know that the majority of highly luminous galaxies in the universe emit the bulk of their energy in the far infrared, and such galaxies may play a key role in galactic evolution.

Schmidt understood the result immediately and wondered how infrared galaxies could be more numerous than quasars. If they were related at all, he conjectured, this meant quasars would have to be obscured for long periods of time. Could this be evidence that luminous infrared galaxies play a dominant role in the formation of quasars? As Donald Lynden-Bell had pointed out years earlier, if quasars were much more plentiful in the early universe, then there must be dormant black holes in many galactic nuclei. But what caused some to blaze forth as quasars, while others remained inert? Perhaps tidal disruption of stars orbiting close to the black hole?

Ultraluminous Infrared Galaxies: Celebrity Rock Stars

Sanders found himself at the nexus of unique observational opportunities as data were becoming available over an increasingly wide range of wavelengths. Following a hunch, he decided to select and analyze what he whimsically termed the "top ten" most luminous infrared galaxies from the IRAS bright galaxy survey.[148] At first, these 10 objects appeared to be nondescript: they did not appear in any of the usual lists of galaxies, such as the NGC catalog or optical spectroscopic surveys. Instead, they were included in Fritz and Margrit Zwicky's 1971 *Catalogue of Selected Compact Galaxies and of Post-eruptive Galaxies*[149] and Chip Arp's *Atlas of Peculiar Galaxies*. Optical counterparts of the host galaxies could be seen on prints of the Palomar Observatory Sky Survey. While Soifer had relied on published classifications of galaxies for his plot of bolometric luminosities, Sanders determined the morphologies from scratch. He imaged each of his 10 objects with the Palomar 60-inch telescope, took spectra with the 200-inch, and identified them with their counterparts on the 48-inch Schmidt using overlays.

From these data, Sanders suggested an evolutionary path leading from infrared-bright galaxies to quasars. In his scenario, the first stage of the transition occurs when gas and dust are driven to the center of an infrared galaxy which contains either a dust-shrouded starburst, an active black hole, or a pair of merging black holes. The galaxies' star-forming regions are embedded in dust, which absorbs ultraviolet and optical radiation. The absorbed energy heats the dust until it radiates at a temperature of 30 to 60 K. Then, in a second stage, powerful winds—created by supernova explosions and radiation pressure—sweep the dust clear of the nuclear region, allowing the embedded quasar to outshine the decaying starburst. A bright "naked" quasar makes its appearance.

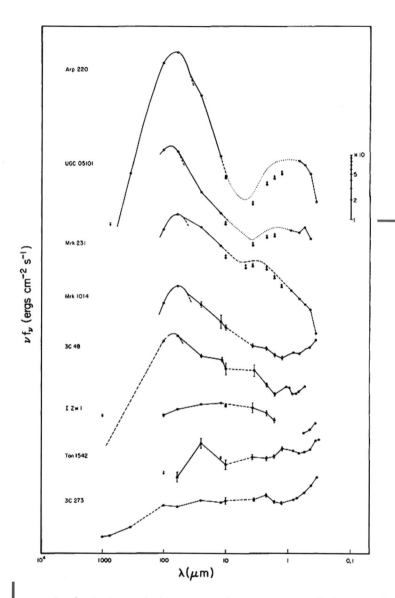

The significant role that dust plays in reshaping the energy distributions of galaxies is revealed in this plot of the flow of energy with wavelength. Here Arp 220—which stood out as the most luminous source during the IRAS survey and was the most luminous ULIRG known in the early 1980s—is compared to other objects, including the well-known quasars 3C 48 and 3C 273. Arp 220's large infrared bump is thought to be due to its quasar nucleus being deeply embedded in dust. The dusty quasar 3C 48—itself highly reddened—is just beginning to shed its cocoon. Quasar 3C 273, at the bottom of the diagram, has shed its dust clouds to reveal its brilliant nucleus, which dominates the spectrum with ultraviolet and X-ray emission. At the time, the long stretch of wavelengths knitted together by the IRAS data in the infrared, Palomar data in the optical, and radio and ultraviolet measurements from other sources was extraordinary. Figure 17 from D. B. Sanders et al., "Ultraluminous Infrared Galaxies and the Origin of Quasars," *Astrophysical Journal* 325 (1988): 74–91, reproduced with permission from D. B. Sanders, University of Hawaii.

Sanders plotted the energy spectrum for each of his 10 ultraluminous infrared galaxies, known as ULIRGs, between 0.44 and 100 microns,[150] using optical and near-infrared photometry from Palomar plus data from the infrared satellite. What he saw on his plot astonished him. The far-infrared emission—beyond 40 microns—dominated the total energy. In addition, when the spectral energy distributions of the ULIRGs were sorted and plotted by increasing luminosity, the enormousness of the far-infrared energy output manifested itself in the form of a prominent bump near 100 microns. Worrying that he might be sticking his neck out too far, Sanders doubled up by finding objects that represented an intermediate stage between the ultraluminous sources and optical quasars. In these transition objects, the central powerhouse was partially visible through its surrounding cloud of dust. These so-called "warm" sources appeared to have blown away part of their gas and dust to reveal the interior cocoon that harbors the starburst. When Sanders published these ideas in a 1988 paper, they were immediately challenged—even laughed at. Yet his arguments were sufficiently convincing to smooth the path to eventual acceptance.

There was a growing appreciation that the more luminous a galaxy is at infrared wavelengths, the more it shows signs of tidal interactions and mergers. The fraction of strongly interacting or merging systems increased from approximately 10% at lower luminosities to close to 100% at the highest luminosities, including in the "top ten" sample. The brightest sources appeared to form stars 10 to 100 times faster than the Milky Way galaxy. From direct images of the 10 ULIRGs, Sanders constructed contour maps. Nine of the ULIRGs turned out to be spatially extended systems. All appeared to be strongly interacting disk galaxies in an advanced stage of merging, as judged from luminous jets, tidal tails, and the 1972 models by Alar and Juri Toomre.

What remained unclear was which combination of mechanisms generated these high luminosities. Sanders's observations at Palomar revealed evidence for both starburst and quasar sources of energy. Quantifying the relative contributions of these two sources of extreme galactic luminosity has kept astronomers occupied for several decades and is currently still under debate.

Chapter 7

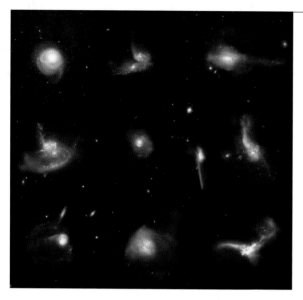
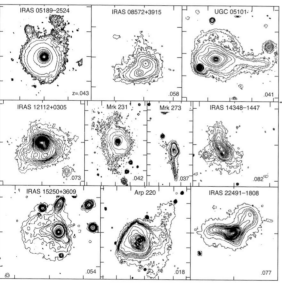

Two mosaics of the "top ten" ultraluminous infrared galaxies and pairs of galaxies display optical direct images (top) and the corresponding contour maps (bottom). These images reveal that nine of the ultraluminous infrared galaxies feature large-scale peculiar structures that include tidal signatures such as tails or bridges. The contour maps detail the morphology of the ULIRG host galaxies. The many signs of tidal perturbations provide strong circumstantial evidence for ongoing mergers of two spiral galaxies. Top: Mosaic kindly made available by Dave Sanders. Bottom: Figure 4 from D. B. Sanders and I. F. Mirabel, "Luminous Infrared Galaxies," *Annual Review of Astronomy and Astrophysics* 34 (1996): 749, reproduced with permission from D. B. Sanders, University of Hawaii.

Cleaning and aluminizing the 200-inch mirror. Photo by the author.

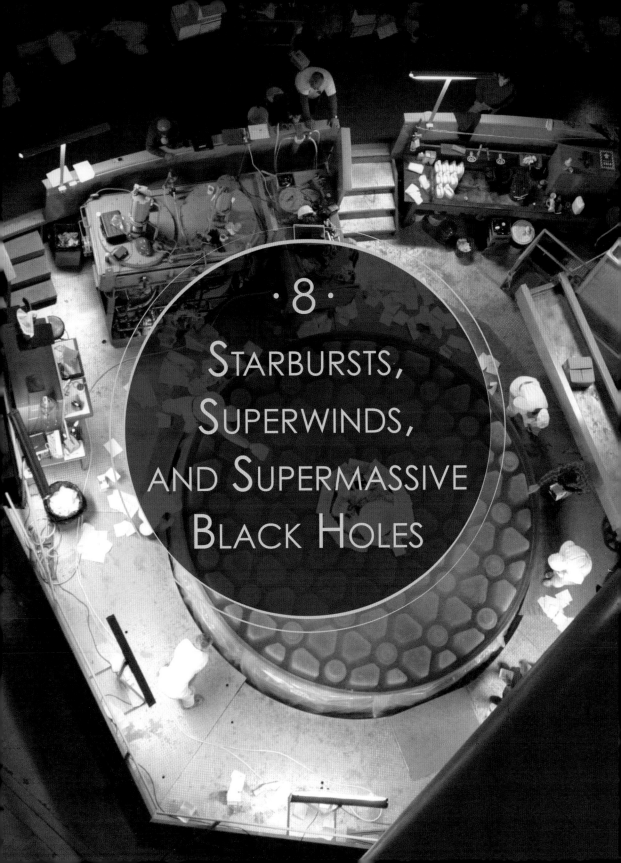

· 8 ·
Starbursts, Superwinds, and Supermassive Black Holes

Chapter 8

Stephan's Quintet, a compact group of galaxies, is a cornucopia of violent events such as collisions, starbursts, supermassive black holes, large-scale shock fronts, and gravitational tides. Confined by mutual gravity, its members have repeatedly engaged in collisions and close passages, some barreling through each other at supersonic speeds and stripping each other of their gas. Unable to survive the collisions *in situ*, the gas lies now suspended in the middle of the group, visible as a long blue arc that exceeds the size of our Milky Way galaxy. The gas in this arc is excited by shocks and heated to 6,000,000 K. In this composite, an X-ray image of the shocked gas is superposed on an optical image of Stephan's Quintet. Since the galaxy to the lower left is in the foreground, there are only four physically associated members shown here, with one member lying outside the image. Credit: X-ray (NASA/CXC/CfA/E. O'Sullivan); optical (Canada-France-Hawaii-Telescope/Coelum).

On the one hand, Tom Matthews's and Allan Sandage's identification of radio "stars" that later turned out to be quasars, and Maarten Schmidt's recognition of their immense distances in 1963, laid the foundation for our growing understanding of supermassive black holes in galaxies. And on the other, Fritz Zwicky's and Chip Arp's first evidence for the importance of galaxy interactions and mergers pointed to what eventually would become the modern notion of hierarchical galaxy assembly. Yet the details, and the realization that galaxies are much more active and interactive than previously thought—not "island universes" evolving only through their internal dynamics—took several decades to fully develop.

FLASHING GALAXIES AND THE PROVENANCE OF HELIUM

Although Zwicky and Arp had recognized the *morphological* distortions of gravitationally interacting galaxies, it took spectroscopic observations by Arp and by Wal Sargent to yield first hints that far more than just morphological changes are taking place. Tidal interactions also dramatically roil the gaseous and stellar contents of galaxies: when the rarefied gas within galaxies is crunched during collisions and mergers, millions—even billions—of luminous young stars can form nearly simultaneously. Astronomers call this event a starburst.

While assembling his *Atlas of Peculiar Galaxies*, Arp noticed that some of the spiral galaxies in his sample had high-surface-brightness companions at the ends of their spiral arms. He took spectra of six such companions with the 200-inch telescope and recognized three kinds of radiation: emission lines from excited gas; a blue continuum of the kind radiated by young O- and B-type stars; and absorption lines from the Balmer series of hydrogen—a hallmark of juvenile A-type stars. Arp inferred that this type of spectrum was likely produced by a recent bout of star formation. In an insightful 1969 paper,[151] he wrote that right after a burst of star formation, massive hot O and B stars radiate a blue continuum. Later, the Balmer absorption lines of less-massive A stars increasingly dominate the galaxy's spectrum for a prolonged period. Although such spectra had also been noticed in blue compact galaxies by Zwicky[152] and by Sargent,[153] Arp appears to have been the first to correctly interpret them as signifying a recent youthful stage of vigorous star formation. He thereby connected these A-type spectra to what today would be labeled a "post-starburst" galaxy. However, the main thrust of his paper was to convince the reader that these bright companions were somehow ejected from their parent galaxies. Other astronomers considered them evidence for rampant *in situ* star formation—the consequence of gas having been dumped onto a companion galaxy during a recent collision.

Sargent was so intrigued by the compact extragalactic objects that his colleague Zwicky had cataloged during a pioneering 20-year-long campaign[154] that he took spectra of 141 of them with the 200-inch telescope. In a 1970 paper, he described how much Zwicky's objects differed from classic elliptical, spiral, and irregular galaxies.[155] Two extreme dwarf galaxies—designated I Zw 18 and II Zw 40 (numbers 18 and 40 in Zwicky's first and second privately circulated lists, respectively)—caught Sargent's attention so forcefully that he began a long collaboration with Carnegie astronomer Leonard Searle with the goal of understanding and modeling them. Since both objects were intrinsically faint, the two astronomers took their data—photographic and photoelectric spectra, photographs, and photoelectric photometry—with the 200-inch telescope.

The cores of the two compact dwarf galaxies, with their blue stellar continua and booming emission lines from ionized gas, resembled the large H II regions that dot the faces of many spiral galaxies.[156] Searle and Sargent estimated that it would take between 1,000 and 10,000 hot O-type stars younger than 10 million years to ionize the volume of gas visible in II Zw 40. Did the presence of so many massive young stars mean that these two dwarf galaxies were young? In the 1970s, astronomers were not yet able to directly observe the oldest stars in such a faint galaxy, so they inferred their existence from the galaxy's color at infrared wavelengths, where red stars dominate. Aiming to measure the amount of light emitted by old stars in II Zw 40, Caltech physicist Gerry Neugebauer observed this object at 1.6 and 2.2 microns, where young blue stars contribute little light. His measurements capped the amount of light emitted by old red stars at 10% of the visual light emitted by the galaxy, with young blue stars supplying the remaining 90%. Thus, blue compact dwarf galaxies were either outright young galaxies or gas-rich systems with very active star formation that resembled the massive H II regions seen in much larger spiral galaxies.

In the 1960s, one central issue in cosmology was the fraction of primordial helium formed during the first few minutes after the big bang as against how much helium had been synthesized since then in stars and distributed via supernova explosions. Searle and Sargent instinctively felt that starbursting dwarf galaxies might yield an answer. If they contained few old stars, and their bursting young stars were a first generation, perhaps their interstellar gas had barely been polluted yet and its composition might be nearly primordial. Indeed, the spectra of I Zw 18 and II Zw 40 showed extremely low abundances of elements heavier than hydrogen and helium. In 1972, Searle and Sargent announced that these two compact dwarfs were the first galaxies known to contain very metal-poor gas and young stars.[157] They argued that the gas in these dwarfs must reflect the primordial abundance of helium, since the abundances of oxygen and neon were much lower than they found in the interstellar gas of the solar neighborhood—while the helium abundance was nearly the same. Relative to hydrogen, the interstellar gas in II Zw 40 contains a mere 10% of the oxygen and neon found in solar-neighborhood gas, and that in I Zw 18 only about 2.5%, the lowest oxygen abundance found until recently.

The discovery of compact dwarf galaxies that rapidly churned out stars, yet showed little chemical enrichment, was a major revelation. Ever since Walter Baade's discovery of two stellar populations, astronomers had associated old stellar populations with low metallicity ("Population II") and young stellar populations with solar metallicity ("Population I"). The discovery by Searle and Sargent exposed a more complex situation. Despite their current intense star formation, these blue compact galaxies had not yet increased their metal abundance through stellar winds and supernova explosions in nearly the same way as had our Milky Way. Thus, these objects appeared to be relatively unevolved, youngish galaxies. But were the dwarf galaxies really young? Were they ancient galaxies producing stars more rapidly now than in the past? Or was their star formation perhaps radically different from that in the solar neighborhood and heavily weighted toward very massive stars?

In a crowning 1973 paper, Searle, Sargent, and William Bagnuolo, then a Caltech graduate student, reasoned that metal-deficient compact dwarf galaxies could not have been *continuously* forming stars that enrich the gas.[158] Continuous star formation at the current level would have consumed—and likely depleted—the observed reservoir of neutral hydrogen gas by now. Therefore, they postulated, the unusually blue colors of the compact dwarfs must be due to *bursts* of star formation. To test their hypothesis, they constructed model galaxies that approximated diverse star formation histories such as extreme youth or episodic bursts of star formation. They approximated these histories by combining aging generations of synthetic star clusters, whose colors they computed from theoretical stellar-evolution tracks. The mix of cluster ages and chemical compositions was then adjusted until the colors of the model galaxies matched the observed colors of the blue compact dwarfs. They found that, while the *light* of dwarf galaxies is dominated by very young stellar populations at the current time, their *stellar mass* appears to be dominated by 10-billion-year-old stars. Therefore, compact dwarf galaxies are likely old galaxies in the process of being rejuvenated through relatively brief, intermittent, and unusually intense episodes of star formation—dubbed "flashing" by Sargent and Searle—separated by much longer periods of quiescence.[159]

Chapter 8

With its blue color, small size, and ongoing starburst, I Zw 18 masqueraded as a very young galaxy of the kind one would expect to find in the early universe—until the ruse was exposed by Searle's and Sargent's exacting observations and models. The most massive hot young stars in I Zw 18 have blasted stellar winds, exploded as supernovae, and blown gargantuan bubbles of gas. Visible as blue filaments surrounding the central region, this gas shines due to the intense ultraviolet radiation emitted by the embedded young stars. Credit: NASA, ESA, and A. Aloisi (Space Telescope Science Institute and European Space Agency, Baltimore, MD).

Since dwarf galaxies are now thought to be the building blocks from which larger systems form through mergers, they play a prominent role in theories describing the formation and evolution of galaxies. Much research has continued beyond the pioneering work of Sargent and Searle, including extensive imaging conducted in part from Palomar. One focus has been whether, due to their collective supernovae-driven winds, these objects are major polluters of the intergalactic medium. However, a further spectroscopic survey led by Sargent for even more extreme objects in Zwicky's catalog found none. In fact, until very recently most later searches for metal-poor galaxies did not exceed the low-abundance record of the archetypal I Zw 18.

Signs of starbursts appear not only in flashing blue dwarf galaxies, but also in many of Arp's peculiar galaxies. In their 1972 landmark paper on "Galactic Bridges and Tails," Alar and Juri Toomre had presented models of four specific systems found in Arp's *Atlas of Peculiar Galaxies* and had hinted that gravitational interactions could funnel gas deep into the centers of galaxies. There the gas would get compressed to densities high enough to trigger bouts of star formation. Two Yale University theoreticians, Richard Larson and Beatrice Tinsley, realized that they could test whether tidal interactions trigger star formation by looking for evidence of starbursts in colliding galaxies. Fortuitously, in 1976 Gérard de Vaucouleurs's *Second Reference Catalogue of Bright Galaxies* had come out with its large compilation of galaxy data, including new photoelectric photometry. Thus, it allowed a systematic and quantitative investigation into how tidal interactions and star formation are linked. Larson and Tinsley plotted two separate empirical graphs, one showing the colors of "normal," relatively isolated galaxies selected from Sandage's 1961 *Hubble Atlas of Galaxies*, the other displaying the colors of peculiar and often gravitationally interacting galaxies selected from Arp's 1966 *Atlas of Peculiar Galaxies*.

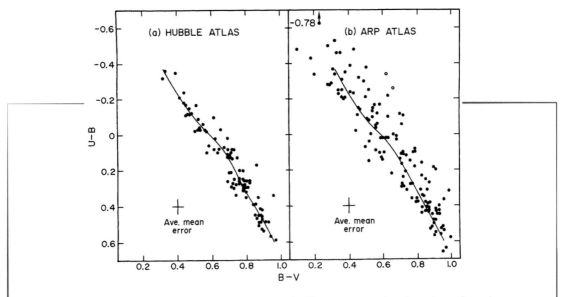

The colors of normal, relatively isolated galaxies (left) are compared to the colors of peculiar, interacting galaxies (right). The interacting galaxies have a larger dispersion in color and tend to be bluer (upper left of the diagram) than the normal, relatively isolated galaxies. From R. B. Larson and B. M. Tinsley, "Star Formation Rates in Normal and Peculiar Galaxies," *Astrophysical Journal* 219 (1978): 46–59, Figure 1. © AAS. Reproduced with permission.

To understand their results, Larson and Tinsley calculated extensive sets of galaxy models, changing parameters such as the strength and duration of starbursts, ranging from sporadic quick bursts to star formation that decreases smoothly over time. To approximate Arp's bluest galaxies with their models, they added young stars to the underlying old stellar population. Because such stars are so bright, it takes only 5–10% of a galaxy's mass in newly formed stars to reproduce the observed blue colors. Their models showed that there was more vigorous star formation in interacting galaxies than in relatively isolated ones. By a process of elimination, Larson and Tinsley concluded that the scatter in the color diagram for the Arp galaxies could only be explained by new or fading bursts of star formation triggered by tidal interactions.

Treading new ground, the two Yale theoreticians forged a causal connection between the bursts of star formation noted by Arp and Sargent and the close tidal interactions and outright collisions between galaxies observed by many. The relatively blue colors, high surface brightness, and gas-rich bodies of many interacting galaxies meant that significant inflows of interstellar gas do occur during tidal interactions. The Toomre brothers had dubbed this process "stoking the furnace." This simple idea gathered much additional observational and theoretical support after Larson and Tinsley's seminal work.

Galactic Bubbles

The first studies of interacting galaxies were stimulated by the objects' strange morphologies. Later came the recognition that some morphological peculiarities may arise from large-scale star formation induced by such interactions. Yet there were also galaxies vigorously forming stars without any apparent signs of interaction with other galaxies. As an example, in the early 1980s astronomers were puzzling over the nature of a pair of extraordinarily bright knots noticed more than three decades earlier lying in an isolated peculiar dwarf galaxy named NGC 1569, also known as Arp 210. Walter Baade first spotted these two knots in 1931 through the murk of the low-galactic-latitude sky, and they were unusual enough to attract attention. Fifty years later, Arp and Allan Sandage—who had been feuding over galaxy redshifts for over a decade—made peace and combined resources to investigate these knots. Safeguarded in a drawer, Sandage had a series of short-exposure photographs taken at the 200-inch telescope by Edwin Hubble in 1953, shortly before his death, and by Walter Baade. The plates were of such high quality as to resolve the amorphous galaxy into regions of ionized hydrogen and individual stars. The question was, were the two bright concentrations near the center of the galaxy single, exceptionally bright supergiant stars? Or were they a new class of "super star clusters"?

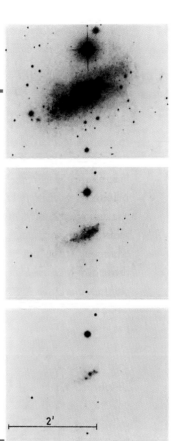

Given the small dynamic range of photographic plates, astronomers routinely took multiple stepped exposures of galaxies to bring out areas of different brightness. The early 1950s photographs of NGC 1569 taken with blue-sensitive emulsions demonstrate such a series. Top, the 25-minute blue exposure taken by Hubble reveals the smooth, bright inner part of the galaxy surrounded by gaseous structure. Below, two one-minute exposures taken by Hubble and Baade, respectively, begin to resolve the two brightest star clusters in the galaxy, as well as some individual stars and H II regions set against the galaxy's bright disk. The fuzzy appearance and slight elongation of the bright knots, in contrast to the sharp, round images of stars in the field, supported the idea that they were extended objects. The overexposed bright object above the galaxy is a foreground star in the Milky Way. Figure 1 from H. Arp and A. Sandage, "Spectra of the Two Brightest Objects in the Amorphous Galaxy NGC 1569—Superluminous Young Star Clusters—or Stars in a Nearby Peculiar Galaxy?," *Astronomical Journal* 90 (1985): 1163–1171. © AAS. Reproduced with permission.

On spectrograms of the knots obtained with the double spectrograph and a new electronic detector at the 200-inch telescope, Arp saw stellar-like hydrogen absorption lines that seemed to be flooded with light from aggregates of very young, high-luminosity hot stars.[160] The Hubble Space Telescope later resolved the two knots into their constituent stars, supporting Arp's and Sandage's conclusion that they were young "super star clusters." Furthermore, the "flooding" turned out to be produced by ultraviolet radiation, stellar winds, and supernovae detritus flowing from the more than 1 million stars in each knot. More recently, X-ray images have revealed bubbles of hot gas thousands of light-years across, filled with the ejecta of supernovae. Since the dwarf galaxy's mass is too low to hold on to the hot gas, this gas expands outward and escapes to pollute and enrich the local universe with its cargo of heavy elements, such as oxygen, silicon, and magnesium.

If dwarf irregular galaxies and other gas-rich systems are loaded with fuel, why is not every such galaxy starbursting? What prompted NGC 1569 to form stars so fiercely that the current rate cannot be sustained for long? Why did it wait 13 billion years to starburst? There are suspicions that, even in this case, a recent interaction may have caused the burst, and this dwarf may also suffer from violence of a more localized nature. After all, small groups of galaxies are known to be violent environments. It is, therefore, interesting that the estimate of the distance to NGC 1569 has recently been revised from 7.8 million light-years to nearly 11 million light-years, a 50% increase. The new distance places NGC 1569 in the middle of a group of about 10 galaxies centered on the spiral galaxy IC 342, making it more likely that gravitational interactions between NGC 1569 and other group members are responsible for shocking its gas and whipping up its starburst. This dwarf galaxy has been producing stars 100 times faster than the Milky Way nearly continuously for the past 100 million years!

Galactic Winds

The "irregular" galaxy Messier 82, a companion to the regular spiral M81, has been studied extensively at optical wavelengths and was one of the first galaxies in which violent processes were recognized. When in 1961 astronomer Roger Lynds of Kitt Peak National Observatory discovered that M82 was also a radio source with an abnormal spectrum, he teamed up with Allan Sandage to take a series of diagnostic photographs with the 200-inch telescope. Using special filters to facilitate the detection of any Hα line emission and star formation, they found a chaotic mixture of luminous filaments and dark dust lanes apparently emanating from M82's center. They also saw spectroscopic evidence for gas moving away from the center of the galaxy vertical to its plane at more than 100 km/s, and perhaps as much as 1,000 km/s. Altogether, the evidence suggested to Lynds and Sandage that around 1.5 million years ago there was an extremely energetic expulsion of gas from the central regions of M82. In a much-cited 1963 paper,[161] they concluded that this might be the first direct evidence of an explosion in the center of a radio galaxy.

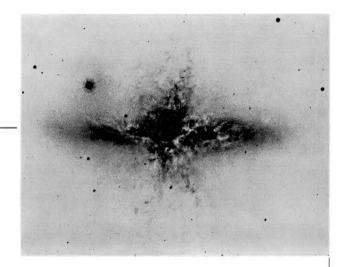

A photograph of Messier 82 shows an extensive system of gas and dust filaments along the minor axis, suggesting that an ancient colossal explosion may have taken place in the nucleus. In 1963—before the age of digital photography—this composite image was made by photographically subtracting a yellow broad-band continuum image from a narrow-band Hα image, both taken with the 200-inch telescope. Figure 8 from C. R. Lynds and A. Sandage, "Evidence for an Explosion in the Center of the Galaxy M82," *Astrophysical Journal* 137 (1963). © AAS. Reproduced with permission.

That claim was followed up shortly thereafter by Sandage and photographic expert William C. Miller, who continued photographing M82 through special filters at the 200-inch telescope and discovered an even vaster system of gaseous filaments extending 4,000 parsecs above and below the plane of the galaxy into the halo of M82. These filaments, glowing in the Hα emission line of hydrogen, seemed to confirm that an explosion had recently occurred. In addition, a high level of polarization suggested that relativistic electrons spiraling in a large-scale magnetic field were emitting synchrotron radiation.

In 1972 Sandage and Carnegie postdoctoral fellow Natarajan Visvanathan returned to studying M82, but this time it became clear that the filaments were polarized due to electron-scattering of light from the galaxy's disk rather than to synchrotron emission. Their observations of Hα polarization weakened what had been considered the best-observed case of an actual explosion in an extragalactic radio source. In a section of their paper entitled "Irreconcilable Problems" the two researchers bemoaned their lack of understanding of the radiation mechanism of the filaments, closing with "Our new observations have made this galaxy a mystery to us."[162]

Chapter 8

Messier 82 is an archetypal starburst galaxy. The apparently cigar-shaped body is a rotating stellar disk viewed nearly edge-on, in which star formation is strongly enhanced. The filaments of gas being ejected from the galaxy's disk trace a galactic superwind driven by many supernovae exploding in the starburst. When a massive young star explodes as a supernova, it creates an expanding bubble of million-degree gas. Whereas in a normal disk galaxy supernovae occur every hundred years or so, in a starburst galaxy they are so frequent and concentrated that their bubbles overlap, creating a large reservoir of hot gas. As this million-degree gas expands, it meets resistance from the dense gas within the plane of the galaxy, but nearly none toward the poles. So the hot-gas reservoir blows out gas and dust perpendicular to the galaxy's disk. The filaments visible in the image mark a superwind of ionized hydrogen, nitrogen, and sulfur escaping the galaxy. Credit: NASA/STScI/SAO.

Over the following decade and through the efforts of many observers, M82's filaments emerged as evidence for a galactic superwind—gas being blown out of the plane of the disk by supernovae exploding in great numbers during a starburst. There is now evidence that fresh gas was dumped onto M82 during a recent interaction with M81, leading to the starburst and subsequent superwind. Even grazing collisions between galaxies can lead to mass transfers between the participants, where the outer gas from one galaxy is dumped into the central region of the other.

Star Formation in the Middle of Nowhere

Galaxy-galaxy interactions are one way to trigger starbursts, as was demonstrated by Larson and Tinsley's work. Galaxy-gas interactions are another way, as can be witnessed in Stephan's Quintet (also known as Arp 319 or Hickson Group 92). Shown at the beginning of this chapter, Stephan's Quintet, a compact group of five galaxies, has intrigued astronomers since its discovery in 1877. The group evolves not only through ongoing gravitational interactions between its members, but also through outright collisions between the group members and the gas that is distributed within the group.

Although a detailed reconstruction of the group's past dynamical history is still in progress, it is thought that a series of previous encounters between group members stripped much of the neutral hydrogen out of their bodies and left it strewn across the group's center. At present, one member of the group, a spiral galaxy that formerly had a gaseous disk, is plunging toward the group's core at around 900 km/s. Along the way, it collided head-on with another group member, triggering in its own disk an expanding ring-shaped shock wave resembling the one observed and modeled in 1976 by Roger Lynds and Alar Toomre.[163] Then, maintaining its breakneck speed, the spiral galaxy struck the hydrogen-rich tidal tail of another group member. The collision is generating an immense and powerful shock wave that stretches over 40 kiloparsecs, substantially larger than the entire Milky Way galaxy. First detected in the early 1970s as a mysterious ridge of radio emission *between* the galaxies—and hitherto invisible in optical light—the shock wave remained unexplained until imaged at other wavelengths. Direct evidence that this ridge is the result of the spiral galaxy plowing into the gaseous debris of another group member was to come from the 200-inch telescope.

First, there were more surprises in store. Kevin Cong Xu became "addicted to Stephan's Quintet" when he was a postdoctoral fellow in the 1990s in Heidelberg.[164] While working on compact groups of galaxies, he and his collaborators discovered an unexpectedly bright infrared source that lay at the northern end of the mysterious radio ridge—and presumed shock wave—in Stephan's Quintet. When they first observed this source, they thought there was something wrong with their instrument. Hidden from optical view by surrounding dust, this source—which they dubbed Source A—appeared to be a starburst. But it was the source's strange location that caught their attention: it was 20 kiloparsecs away from the nearest galaxy, in the middle of nowhere. What extreme, implausible set of circumstances would trigger a starburst outside the boundaries of a galaxy?

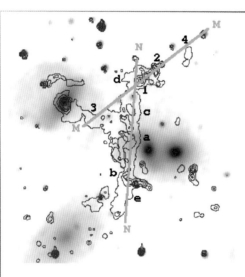

Optical spectra of Stephan's Quintet helped clinch the dynamical relationship between the long intergalactic shock front and the isolated starburst known as Source A. The schematic shows the two placements, M and N, of the slit of the double spectrograph of the 200-inch telescope across Stephan's Quintet. In Position M the slit crossed both the shock region (red contours) and Source A (located at Position 1), yielding their relative velocities and turbulence, while in Position N the slit was aligned along the shock region. In this schematic, the contours mark Hα and [N II] line emission from ionized gas: blue signifies emission from low-velocity gas (about 5,700 km/s), while red marks emission from high-velocity gas (6,600 km/s), including the shock front. The shock has by now been detected in X-rays, radio, Hα, and molecular hydrogen lines. Figure 6 from C. K. Xu et al., "Physical Conditions and Star Formation Activity in the Intragroup Medium of Stephan's Quintet," *Astrophysical Journal* 595 (2003): 665–684. © AAS. Reproduced with permission.

After joining the Infrared Processing and Analysis Center at Caltech, Xu turned to the 200-inch telescope to investigate whether these two highly unusual phenomena—the isolated starburst and the large-scale shock front—were physically related and perhaps even dynamically linked. Spectral emission line ratios measured in the gas surrounding Source A confirmed that the high-velocity impact of the spiral galaxy, visible just west (to the right) of the vertical shock front, had caused both the starburst in Source A and the large-scale shock. Since the emission lines from the ionized gas were also extraordinarily broad—normally a sign of extreme turbulence—he was confident of the cause and effect.[165]

The violent collision of a galaxy with gas filaments in Stephan's Quintet is distinct from a mere tidal interaction, where two galaxies pass by each other in a gentler fashion. This is perhaps the first time a starburst was found to have been induced by a collision between the gas in a high-speed galaxy and the intragroup medium. The unexpectedly strong emission of Hα from Source A seen in the Palomar spectra turned out to be more powerful than the total energy emitted in diffuse X-rays. This finding was exciting enough to prompt Xu and colleague Phil Appleton to point the Spitzer Space Telescope's far-infrared-wavelength detector at the center of the shock region. There they discovered other, even stronger lines of molecular hydrogen, the most direct evidence yet for the fuel on which star formation depends. Yet there is almost no star formation within the galaxy members, and also little outside. The "starbursting" Source A produces only around three-quarters of a solar mass per year in new stars, a small amount when compared to starbursting galaxies. Thus, Stephan's Quintet remains somewhat enigmatic, fondly referred to by some as a "turbulent mess."[166]

Maelstrom in the Nucleus

Palomar astronomers had had the good luck of detecting quasar host galaxies, acquiring observations in support of galactic collisions triggering nuclear activity, and recognizing major bursts of star formation. Then, in 1983, data pouring in from the just-launched Infrared Astronomy Satellite (IRAS) provided a stunning new view of the dust-enshrouded centers of galaxies. Follow-up observations showed that some optically inconspicuous galaxies, such as Arp 220, could emit as much as 99% of their energy at infrared wavelengths. Whatever the sources behind such powerful radiation were, evidence was building that the more luminous a galaxy is in the infrared, the more likely that it is in the process of colliding and merging.

On August 17, 1989, a clear night with good, 1-arcsecond seeing, UC Berkeley astronomer James Graham and his colleagues observed Arp 220 with the 200-inch telescope equipped with an infrared camera and a 2.2-micron photometric filter. They alternated between 5-second integrations of Arp 220 and integrations of a nearby star of comparable brightness, and repeated the sequence several times. After subtracting the thermal emission from the sky and telescope, they applied a clever "shift-and-add" technique to correct for first-order blurring of the image due to seeing. Their exacting techniques resolved a hidden double nucleus in Arp 220: two distinct infrared sources about 1-arcsecond apart, which corresponds to about 1,000 light-years in projection. The

CHAPTER 8

discovery of the two close objects—each then thought to contain a dust-enshrouded Seyfert nucleus or quasar—along with nonconcentric infrared isophotes, a disturbed optical morphology, and remnants of tidal tails, provided ample evidence that Arp 220 is a merger in its final stages. They announced their discovery in 1990 in an *Astrophysical Journal* paper.[167]

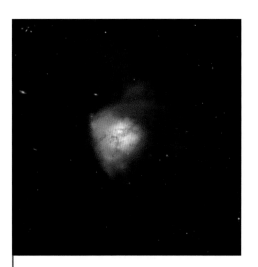
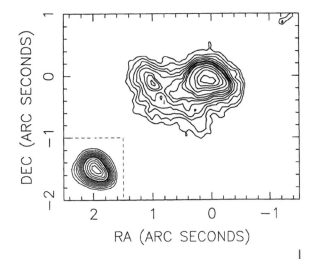

In Chip Arp's *Atlas of Peculiar Galaxies*, Object 220 (Arp 220) appears as a puny faint nondescript galaxy with adjacent loops. A dust lane bisecting the galaxy appears in the optical, but there is no hint from a blue-light photograph of the maelstrom caused by the two merging, X-ray-emitting nuclei in its core. Arp 220, the nearest ultraluminous infrared galaxy, has the highest bolometric luminosity in the local universe, churning out radiation at the rate of 6×10^{45} ergs/sec—more than a trillion times higher than the sun. Yet it is a true enigma, since its primary energy source remains hidden behind approximately 10 magnitudes of extinction, even at 2.2 microns. It is thought that a collision between two spiral galaxies occurred several hundred million years ago, triggering widespread starbursts that have now shrunk to a relatively small region. Left image, spanning 84,000 light-years across: A Hubble Space Telescope image of Arp 220 at visual wavelengths shows the faint remnants of tidal tails, signs of a nearly complete merger. But there is no sign of a double nucleus, only a central region crisscrossed with dust and peppered with knots of star formation. Right image, 4,000 light-years across: A contour map of the region surrounding the core of Arp 220, based on infrared imaging at the 200-inch telescope, clearly shows the two nuclei. In the lower left-hand corner are the contours of a star observed at the same time. Left: NASA, ESA, the Hubble Heritage Team (STScI/AURA)-ESA/Hubble Collaboration, and A. Evans (University of Virginia, Charlottesville/NRAO/Stony Brook University). Right: Figure 1 from J. R. Graham et al., "The Double Nucleus of Arp 220 Unveiled," *Astrophysical Journal* 354 (1990): L5–L8. © AAS. Reproduced with permission.

Infrared luminous galaxies such as Arp 220 allow astronomers to study the balance between two fundamental modes of central energy generation. The first mode is thermonuclear burning—the conversion of hydrogen to helium and heavier elements—in millions of stars formed in a central starburst. The second mode is the release of gravitational energy by material falling toward a supermassive black hole and frictional heating in its accretion disk of gas. It now appears that the higher the infrared luminosity of a luminous galaxy, the more it tends to be driven by a supermassive black hole (or holes); and the lower the luminosity, the more it tends to be driven by starbursts. Both forms of energy release can produce massive outflows of gas.

Black Holes: Central Monsters

The discovery of quasars in 1963, and especially their small intrinsic sizes, had led to the hypothesis that these extremely powerful sources derive their energy from the accretion of material onto compact, extremely massive objects, most likely central black holes. At the time, it was believed that black holes are rare and exotic objects inhabiting only a small number of galaxies. Yet intensive follow-up observations by Maarten Schmidt during the 1960s and early 1970s suggested that quasars were much more numerous—by a factor of perhaps 1,000—around redshift 2, when the universe was much younger than at present. That raised an interesting question: If most quasars have died out by now, shouldn't there be glowing embers left over from their brilliant past? If so, dead quasars might be hiding in the centers of nearby galaxies in the form of supermassive black holes.[168]

But how would we know? Evidence for black holes is, by nature, circumstantial and indirect. For supermassive black holes, it is most often acquired from their effects on the dynamics of the surrounding stars and gas that form the visible galactic nucleus. For example, a central hole's strong gravitational pull accelerates stars in their orbits as they pass near to its sphere of influence. This is observable as a sharp increase in the light and in the local stellar velocity dispersion. Often barely discernible with ground-based telescopes, these effects are easiest to probe in nearby galaxies with quiet nuclei rather than in those with raging quasars at their centers. In the latter, the fireworks of starbursts and accretion-disk activity tend to drown out weaker signatures from stars and gas dominated by gravitational forces.

To Wallace Sargent and Caltech graduate student Peter Young, the well-known giant elliptical galaxy M87—a host to powerful radio and X-ray sources—appeared a promising candidate for the first detection of a supermassive black hole. Sitting at the center of the Virgo Cluster about 54 million light-years from Earth, M87 has a small, abnormally bright nucleus from which emerges a jet of relativistic plasma. Sargent and Young took a two-pronged approach to exploring evidence for any central black hole.

First, in 1977 Young, Jim Westphal, and several collaborators made the first accurate measurement of the surface brightness of the core of M87 using new photoelectric detectors (including a first CCD) mounted on the 60-inch and 200-inch telescopes. They described the light profiles in the central region as "grossly deviant," with a central excess of luminosity—a "cusp"—not seen in similar data for nearby normal elliptical galaxies and not predicted by any standard models.[169] The team cautiously concluded that a black hole of around 3 billion solar masses would fit the observed profiles.

Second, British astronomer Alexander Boksenberg's much-heralded two-dimensional photon-counting detector was flown to Palomar and to Kitt Peak National Observatory to help measure the orbital velocities, velocity dispersions, and line strengths of stars across the face of M87. For comparison, they also measured NGC 3379, a normal elliptical galaxy. The new detector was sensitive enough for even the outer parts of the galaxies where the surface brightness was much below that of the night sky. Although individual stars could not be discerned at the galaxies' distances, their organized and random motions were measured from absorption line redshifts and the breadth of the spectral lines.

The spectra were analyzed by Young, who measured the velocity dispersions of the stars (the spread of random motions) along the axes of M87 and NGC 3379. He found that the central regions of the two galaxies differ dynamically: near the center of M87, the stars move faster and faster, with random velocities peaking at 300 to 500 km/s, while in NGC 3379 no such central velocity spike was observed. Thus, in M87 a good fit of the stellar-dynamical models to the observations required adding a dark central object of around 5 billion solar masses.

While the observations did not prove that the massive object in M87 was indeed a black hole, they were strongly suggestive and "the most attractive of the models considered." Further supporting their hypothesis, Sargent and Young calculated that a central black hole of 5 billion solar masses could easily fuel itself by consuming mass lost from stars in the cusp. In so doing, it could produce the prodigious energy output of 10^{42} ergs per second measured in Virgo A, the radio source corresponding to M87. Eventually, the Hubble Space Telescope revealed a rapidly rotating small gas disk at the very center of M87. The central mass determining its rotation was estimated to be more than 2 billion solar masses, roughly comparable to Sargent and Young's early estimate.

One astronomer not convinced that Sargent and Young's findings could be generalized was Alan Dressler of the Carnegie Observatories, across town from Caltech. He argued that finding evidence for a black hole in Messier 87 at the distance of 50 million light-years had been feasible only because that black hole turned out to be so very massive. So why not look for evidence of black holes in galaxies closer to home? He chose to investigate NGC 1068 and NGC 4151, the two best-known Seyfert galaxies with bright nuclei. Biding his time until the seeing was around one arcsecond at Palomar, he pulled the coordinates out of his briefcase and climbed into the 200-inch telescope's Cassegrain cage, where the double spectrograph was installed. Since no one had yet obtained spectra of galaxies at near-infrared wavelengths, Dressler observed two well-studied nearby quiescent galaxies for comparison: Messier 31 and its elliptical companion M32. Although neither M31 nor M32 showed signs of nuclear activity similar to M87's, they are 20 times closer to Earth, permitting more detailed measurements of their light profiles and stellar kinematics.

> **SIZE MATTERS:**
> In black hole jargon, the surface that bounds the region from which nothing can escape the gravitational pull of a black hole is dubbed the event horizon. The size of this horizon surrounding a nonrotating, uncharged black hole, known as its Schwarzschild radius (R_s), depends on the black hole's mass. For a one-solar-mass black hole, the Schwarzschild radius is about 3 km. For a one-billion-solar-mass black hole, it is 3 billion km. At the distance of the Virgo Cluster of galaxies, the Schwarzschild radius of such a supermassive black hole subtends an angle of only one-millionth of an arcsecond, which is 100 times smaller than what could be resolved in the 1980s with the world's best interferometers.

Examining his data back in Pasadena, Dressler was struck by how much M31's stellar rotation curve stood out. As it crossed the galaxy's nucleus, it went into a complete reversal, whose profile he characterized as "a whoot whoot whoot,"[170] while the stellar velocity dispersion (not shown in the figure) rose toward the nucleus much as had been observed in M87. From the velocity dispersions, he inferred a dark central mass of about 30–70 million solar masses in M31 and 8 million solar masses in M32.[171] Since not as much light was coming from the galaxies' centers as expected for a normal mix of stars, supermassive black holes were—again—the likely culprits.

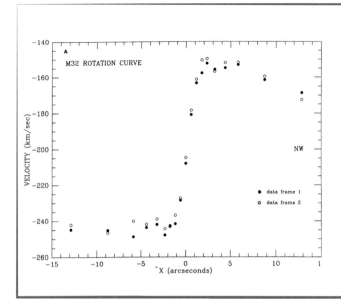

Radial velocities measured from three separate exposures are plotted against distance from the nucleus along the major axis of M31's disk. The nucleus of the galaxy lies at 0 arcseconds on the horizontal axis. The steepness of this "rotation curve" as it crosses the nucleus shows that the nucleus itself is rotating rapidly. Figure 2 from A. Dressler and D. O. Richstone, "Stellar Dynamics in the Nuclei of M31 and M32—Evidence for Massive Black Holes?," *Astrophysical Journal* 324 (1988): 701–713. Reproduced by permission of Alan Dressler, Carnegie Observatories. © AAS.

Dressler sent his data to Douglas Richstone, an expert in modeling stellar orbits at the University of Michigan, asking him whether the data could be interpreted as stars moving in a kind of raceway around a massive dark object. Richstone's answer was yes, which made the strongest case yet for the presence of supermassive black holes in the two galaxies, a result they published together in 1988.[172] Strikingly, these two nearby galaxies—one a small compact elliptical and the other a 100-times-more-luminous spiral—displayed similar photometric and kinematic evidence pointing to supermassive black holes at their centers. In contrast, as far as evidence for such black holes goes, Dressler's original Seyfert galaxy targets were a bust. Serendipity was at work.

The Key to the Vault

Might, then, supermassive black holes be a common feature at the center of most, or even all, galaxies? Having found evidence for a supermassive black hole in M87, Wal Sargent discussed the object with Alexei Filippenko, one of his graduate students. Curious about what fraction of all nearby galaxies in the local universe show signs of nuclear activity, Filippenko realized that more examples and higher-quality data were needed for meaningful statistics. Setting a relatively bright 12.5-magnitude limit, he selected 500 galaxies across all luminosities and Hubble types from the *Revised Shapley-Ames Catalog of Bright Galaxies*. In 1982, Filippenko began a herculean survey with Sargent at Palomar, taking high-signal-to-noise, optical-range spectra of the nuclear region of each galaxy. They used the tried-and-true double spectrograph mounted on the 200-inch telescope, and a 2-arcsecond-wide slit helped isolate the nucleus of each galaxy.

In the early days of this survey, there was no automatic guider at the 200-inch telescope to track the celestial target. Instead, Filippenko or Sargent would push the north-south-east-west buttons on the handheld guide paddle hundreds of times during each night. Although the double spectrograph could be rotated to align the slit with the desired direction across a galaxy from the comfort of the data room, this was not the case for adjusting the grating. Each afternoon, one or the other observer would climb up into the Cassegrain cage to rotate the grating for the spectral range to be measured that night. To that purpose, the spectrograph was bolted to a large rotator ring at the bottom of the telescope. As communication was through walkie-talkies, a voice would reverberate from the data room through the darkened dome: "Turn the dial to the left!" After a few moments, a plaintive voice would respond from the cage: "Which left? Your left? My left?" While observing in the data room, Sargent—widely considered somewhat eccentric—would share his vast extracurricular interests in music, baseball, and sumo wrestling.

The observing continued for nearly a decade and required the allocation of more than 100 dark nights on the Hale telescope. Yet Filippenko's survey remained the largest database of high-quality, homogeneous optical spectra of nearby galaxies for long thereafter. In 1985, Filippenko and Sargent published some initial results from a subset of 35 galaxies in the survey.[173] They had found that a substantial fraction of nearby bright galaxies exhibit signs of activity similar to—but more mild than—that seen in Seyfert galaxies and quasars. Supermassive black holes were likely present, but a definitive result awaited a thorough analysis of the entire sample of 500 galaxies.

> **CRUISING THE CATWALK:**
> Late in the afternoon, before dinner at the Monastery, Wal Sargent would survey the sky and Observatory grounds from the raised catwalk encircling the 200-inch dome. Looking at the sky, he would try to predict how the seeing and clarity would be during the night. Once, several tourists spotted him and asked: "Hey you, how'd you get up there?" In his idiosyncratic way, Sargent looked down at them and growled: "With 30 years of bloody hard work!"
> (A. V. Filippenko, private communication, 2019)

For several years, most of the data from the Palomar spectroscopic survey of nearby galaxies lay fallow in Filippenko's storage room in Campbell Hall on the UC Berkeley campus. It was a giant scientific data mine and a daunting sight: rows and rows of large, old-fashioned 8-track magnetic tapes stacked on five racks, each tape filled with recorded data. One day in the early 1990s, Filippenko—by now an overtaxed Berkeley professor whose personal research had shifted toward supernovae—gave his graduate student Luis Ho the key to the room. It was a momentous occasion, and one that left Ho feeling overwhelmed. Yet the lure of hidden treasures kept him working for three and a half years to upload the data from the disintegrating tapes to various computers.[174] He faced a managerial challenge in organizing the massive database—not very glamorous work. By the end of the process, the tape drives themselves were failing, and it became a desperate race between old and new technology.

Although Ho did not take part in the original data acquisition, he participated in many follow-up observing runs. Soon he realized that good data alone would not suffice: a thorough analysis was needed to tease out evidence that would clinch the case for any potential black holes. One major obstacle to isolating and measuring faint emission lines in galaxy nuclei is the vast amount of light—amounting to 90–95% of the total—contributed by star-forming regions and old stars that populate the nuclear region. Fortunately, the emission line profiles of the fast-moving gas close to a black hole are broadened and therefore distinguishable from the narrow lines emitted by star-forming regions. To remove any distortions from the broad emission lines of interest, Ho designed a custom template to subtract the light of old stars.

When their observations were fully analyzed, Ho, Filippenko, and Sargent deduced that supermassive black holes are quite common in galaxies, especially in elliptical and bulge-dominated types. In 1999, Ho argued that supermassive black holes are a normal component of galactic structure, one that arises naturally in the course of galaxy formation and evolution.[175] Thus, supermassive black holes could no longer be considered rare freaks of nature—intrinsically interesting, but not relevant to the rest of the universe. By providing considerable insight into the nature of nuclear activity in nearby galaxies, the Palomar spectroscopic survey of 500 nearby galaxies set the stage for the next decade.

In 1981, the husband-and-wife team of Manuel Peimbert and Silvia Torres-Peimbert[176] had announced evidence for a supermassive black hole in the nucleus of the nearby galaxy M81.[177] In fact, their announcement had provided Filippenko with much of the initial inspiration for his survey. Seven years later, one of the early results of the Palomar spectroscopic survey was the confirmation of M81's supermassive black hole. Since the random motions of stars at the centers of giant spiral galaxies rarely exceed

200–300 km/s, their detection of broad widths that indicated much higher velocities in M81's central gas had set off an alarm. Such broad-winged spectral lines are a telltale signature of massive black holes, similar to what is seen in Seyfert galaxies and in much more luminous quasars. Only a high concentration of mass—likely a black hole in the center of this otherwise quiescent galaxy—could explain why the gas is moving so fast. In 1996, Ho, Filippenko, and Sargent analyzed the width of M81's broad emission lines and estimated that its central dark object was between 700,000 and 3 million solar masses.[178] Thus, it has an active galactic nucleus, although dozens of times weaker than a Seyfert galaxy and thousands of times weaker than a quasar.

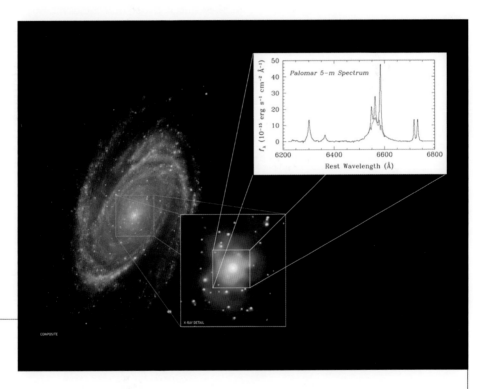

This image of M81 is a composite of infrared and ultraviolet light, while the blue zoom shows the center of the galaxy in X-rays. The second zoom displays the emission line spectrum of the nucleus of M81 around the Hα line after contamination by starlight was removed. The red line marks the broad-winged component of Hα emission, which indicates that gas is moving at extremely high velocities of up to 3,000 km/s. The figure, kindly provided by Luis Ho, is composed of images from the Spitzer Space Telescope (infrared), GALEX (ultraviolet), and the Chandra Space Observatory (X-ray).

Chapter 8

As an outgrowth of the spectroscopic work done at Palomar and other observatories, a team of astronomers using the Hubble Space Telescope discovered that supermassive black holes almost invariably have a mass about 0.4% of the bulge mass of the galaxy in which they are embedded. That is, the mass of a central black hole strongly correlates with the mass of the bulge of its host galaxy. Since the bulge mass is inferred from the stellar velocity dispersion, a more directly observable correlation exists between the black-hole mass and the velocity dispersion measured in the bulge. But what mechanism might intimately connect a solar-system-sized black hole at the center of a galaxy to the large-scale properties of the galaxy, and vice versa?

One way to deliver gas to the central black hole quickly and efficiently is through galaxy mergers. Merging galaxies induce tidal torques and shocks that remove angular momentum from the gas and allow it to fall toward their nuclei. Some of the gas is converted into stars via starbursts. A smaller fraction of the gas dribbles down to the black hole and stimulates nuclear activity. Since most galaxies contain black holes, when two galaxies merge, so do their black holes. In the extreme, accretion disks around these supermassive black holes can manifest themselves as quasars and outshine their entire host galaxy.

Recent observational evidence suggests that quasars can also drive gas outflows (winds). These outflows, often stronger and faster than those generated by starbursts, may rather abruptly end the concomitant starbursts, thus "quenching" star formation and leading to so-called "red and dead" ellipticals. This idea remains controversial. Yet it all seems to hang together: galaxy interactions and mergers drive gas to their centers, feeding both starbursts and active galactic nuclei (AGNs) to varying degrees, and both the starbursts and AGNs drive gas outflows that quench the starburst and diminish the fuel available for further AGN activity. This *may* one day explain the close link between the masses of bulges and those of the central black holes.

The honeycomb mirror. Photo by the author.

· 9 ·
Probing the Gaseous Universe

Chapter 9

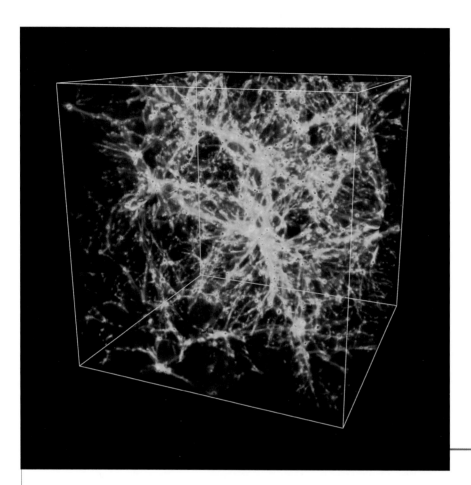

One of the pillars on which the big bang theory rests is the census of baryons, the common matter of which stars, planets, and we humans are made. Yet, when astronomers set out to add up the baryon content of the local, low-redshift universe, the numbers fell 30–40% short of the value predicted by the big bang theory. The remaining baryons have not been convincingly observed as yet, but are thought to exist in a vast web of hot gaseous filaments dubbed the "warm-hot intergalactic medium," or WHIM. Computer models predict that such filaments range in temperature from 100,000 to 10,000,000 K and are heated by gravitationally driven shocks. Modeling the gas's behavior in a cube about 400 million light-years across (shown here) yields a gas distribution resembling the structure of a cosmic web. Green regions represent the lowest gas densities within the web, while orange and red regions represent the highest gas densities and sites of galaxy formation. Recent observations in the ultraviolet and in X-rays are just beginning to uncover evidence for the existence of this hypothesized WHIM. Figure 2 from R. Cen and J. Ostriker, "Where Are the Baryons? II. Feedback Effects," *Astrophysical Journal* 650 (2006): 560–572.

In the first two decades of its operation, the 200-inch telescope was used by astronomers to obtain photographs, spectra, and photometric measurements of objects ranging from planets to stars, nebulae, galaxies, and clusters of galaxies. Most of these objects were luminous and directly visible in the eyepiece of the telescope. Within our own Galaxy, astronomers also found evidence for the presence of interstellar material in the form of dark clouds obscuring the luminous band of the Milky Way, absorption lines seen in the spectra of stars, and a general reddening of light from distant stars. However, no similar signs of light extinction had as yet been detected in the spectra of distant galaxies. In fact, a famous spectroscopic study of 800 galaxies published in 1956 concluded—from the lack of intergalactic absorption lines—that space was extremely transparent as far as astronomers could see. This optimistic assessment held for about a decade. Then signs of trouble began to appear in the form of mysterious, knife-sharp absorption lines in the spectra of some bright quasars. Unknown to astronomers at the time, these absorption lines signaled that filaments of nearly invisible gas *do* pervade the universe—our first glimpse of the cosmic web.

A Spike of Light

Maarten Schmidt, who had astonished the world in 1963 by showing quasars to be a new class of enormously energetic extragalactic objects, was poised to reveal another new class of phenomena, even if unwittingly. While searching for more and more distant quasars, he reached unexplored spectral territory as the quasars' higher recession velocities shifted spectral lines from their rest-frame far-ultraviolet into the blue-visual region that matched the sensitivity of Schmidt's photographic plates.

To help identify unfamiliar spectral lines, Schmidt relied on a search list of 37 standard ultraviolet transitions compiled for him by renowned Mount Wilson spectroscopist Ira Bowen. The list included the rest wavelengths of lines already observed in quasi-stellar sources, radio galaxies, the sun's ultraviolet, and planetary nebulae, the latter including predicted as well as observed ultraviolet lines. As neither quasars nor their savage physical environments were yet understood, Schmidt gingerly worked his way "hand over hand" to identify new lines as they appeared in increasingly distant sources.

The spectrum of an object designated 3C 9—the ninth object in the Third Cambridge Catalogue of Radio Sources—caught him by surprise: a towering spike of emission jutted from the far blue end of the recorded spectrum. Schmidt knew that high-energy cosmic objects emit the bulk of their energy in the ultraviolet and that their radiation often involves transitions between the lowest energy states of the hydrogen atom. Still, he repeated the five-hour-long exposure three times until he was confident in his line identification. The spike was made of Lyman-α photons emitted as hydrogen atoms drop from the first excited state to the ground state. Given the universal abundance of hydrogen, such photons were expected to be prevalent throughout the universe.

Discovered by Harvard University physicist Theodore Lyman in 1906, the Lyman-α line lies in the ultraviolet region of the electromagnetic spectrum at a wavelength of 1,216 angstroms. Schmidt calculated that, to be recorded at the far-blue end of his photographic plate, at 3,666 angstroms, 3C 9's Lyman-α line had to be redshifted by a factor of three.[179] At the corresponding redshift of 2.01, the light would have left the quasar when the universe was one-fifth its current age—long before the sun formed and about when the Milky Way galaxy had begun assembling. The light would have left the quasar nearly 11 billion years before striking Schmidt's photographic plate.

Quasars were still an exceedingly mysterious concept when Schmidt exhibited five newly discovered objects in the early spring of 1965 during a colloquium at Caltech. He dramatically unveiled his treasures one by one with increasing redshift, culminating with 3C 9 and its spike of Lyman-α. In the audience, two clever fourth-year graduate students, Jim Gunn and Bruce Peterson, found their attention riveted not by the hallowed Lyman-α line itself, but by the shape and strength of the wavelength region *shortward* of it. Both independently realized that if the intergalactic medium was filled with neutral hydrogen gas, the atoms situated between the quasar and Earth would absorb the quasar's continuum radiation blueward of its Lyman-α emission line. The cumulative effect of such incremental absorptions, over billions of light-years, would carve out a smooth absorption trough beginning at the quasar's Lyman-α line and extending to shorter wavelengths. The appearance and depth of such a trough could put stringent upper limits on the amount of neutral hydrogen between the quasar and us. In 3C 9, the region in the spectrum shortward of the line should have been completely dark. It wasn't. It was gray, which meant that some light was getting through. Either there was little hydrogen in intergalactic space, or it was all ionized, perhaps by starlight, and incapable of absorbing Lyman-α photons. From the shade of gray, the two graduate students estimated that the quasar continuum was depressed by about 40%.

Within months of Schmidt's colloquium, they had borrowed and analyzed his photographic spectrum of 3C 9, worked out the details and mathematical equations, and published a fundamental paper in *The Astrophysical Journal* entitled "On the Density of Neutral Hydrogen in Intergalactic Space."[180] Their calculations showed that it would take at least one neutral hydrogen atom for every 100,000 ionized hydrogen atoms in intergalactic space to block the quasar's light. The limit seemed small to them—about five orders of magnitude below the expected cosmic density. Since only a weak depression—not a deep trough—was observed in the spectrum of 3C 9, Gunn and Peterson reasoned that either most hydrogen in intergalactic space had already condensed into galaxies, or most of it was ionized and thus rendered incapable of absorbing photons. In the latter case, the epoch during which the universe was filled with *neutral* hydrogen—necessary for the formation of stars and galaxies—had to have been relatively brief. Therefore, they reasoned, neutral hydrogen from the early universe should be detectable in the spectra of objects at high enough redshift. In the final section of their paper, they lamented that "the interpretation of data from very distant sources is no longer a straightforward application of the usual simple cosmological tests that involve only … knowledge of intrinsic source properties." Once more, the universe had manifested its complexity.

This fundamental and highly cited paper led to much subsequent work on the ionization of the intergalactic medium at high redshifts. Although the detection of redshifted Lyman-α emission stirred others to explore the contents of intergalactic space silhouetted against quasars, Schmidt did not succumb to the temptation to change focus. He felt that his first priority was to document the spectral characteristics of quasars themselves, and acquiring sufficient data for the necessary statistics drew him along that path. He searched for quasars at ever higher redshifts, staying far ahead of galaxies in terms of redshift records.

Meanwhile, the quest to map the "Gunn-Peterson trough" began in earnest in 1965 when John Bahcall, then at Caltech, and Edwin Salpeter at Cornell University quickly refined the calculations of Gunn and Peterson. While Gunn and Peterson's smooth distribution of neutral hydrogen gas would carve out a smooth absorption trough, in reality Bahcall and Salpeter envisioned clumps of gas. Instead of a smooth trough, there would be a series of sharp absorption lines, each line produced by an intervening gas cloud, galaxy, or cluster of galaxies.[181] Indeed, as the resolution of quasar spectra improved, astronomers witnessed the trough breaking up into individual absorption lines.

At the time, observers, including Schmidt, struggled to measure redshifts of quasars, often stymied by their complex absorption and emission line spectra. Many of the lines remained unidentified, were unresolved, or displayed complex structure, such as splitting. Sometimes redshifts measured for the same quasar by various observers didn't agree. For example, in 1967 Chip Arp and colleagues identified the blue starlike counterpart to the radio source Parkes 0237-23, a quasar, on 48-inch POSS plates, then took spectra of it at the prime focus of the 200-inch telescope. Although their redshift of 2.22 was based on numerous absorption lines, it inexplicably disagreed with Margaret Burbidge's redshift of 1.95 measured at another observatory.

Hoping to resolve this apparent contradiction, Jesse Greenstein observed Parkes 0237-23 and recorded more than a hundred sharp and deep absorption lines with the prime-focus spectrograph of the 200-inch telescope—and found that *both* velocities were present. An explanation happily occurred to him and Schmidt as they sat together on a plane to discuss their findings while traveling to a conference. Although the quasar itself could have only one emission line redshift, there was no reason why multiple *absorbing* sources couldn't simultaneously be present between the quasar and the telescope—each with a different redshift.[182] Four years later, Roger Lynds of Kitt Peak National Observatory figured out that nearly all the narrow absorption lines in quasar spectra were due to the same transition, the absorption of a photon of 1,216 angstroms, which is Lyman-α.[183] This kicks an electron from the first to the second orbital in the hydrogen atom. Still, were the lines formed in individual gas clouds? Were they intergalactic, in intervening galaxies, in clusters of galaxies, or associated with the quasar itself?

The tools available to astronomers at the time—low-resolution spectrographs, image tubes, and nonlinear photographic plates—were not up to the task of detecting and resolving feeble and crowded absorption lines in the spectra of faint objects. But news of a revolutionary digital detector that could count individual photons offered hope. The new instrument was said to have enough power to wring data from even a faint smudge in the sky, and enough wavelength coverage, from 3,220 to 9,000 angstroms, to record Lyman-α absorption over a wide range of redshifts.

PERFECT INSTRUMENT MEETS PERFECT TELESCOPE

The last thing British physicist Alexander Boksenberg wanted to do was astronomy. His PhD was in atomic physics, firing electrons at atoms. However, the only job he could find upon graduation was in space astronomy, working on rockets and instruments. It turned out to be a fortuitous snag. The idea for a "perfect observing system" for faint objects occurred to him in the late 1960s, while he was leisurely soaking in the bathtub. He wondered, why not "turn your photons to look as big as elephants and then simply count them as they arrive on Earth?"[184] His construct would work somewhat like the human eye, where incoming light impinges on the high-sensitivity retina, which processes the data-rich image before it travels along the optic nerve to be interpreted by the brain. Boksenberg and a crack team of electronics and software wizards emulated biological processes with available hardware to detect, amplify, process, and accumulate photons in real time. Instead of the single-channel detectors then in use, their instrument created two-dimensional images. Dubbed the Image-tube Photon Counting System or IPCS, this instrument is still well-known among older astronomers.

The "retina" of the IPCS was a four-stage high-gain image tube fabricated by EMI, a company known for its recordings, television, and image intensifiers—and famous for having recorded The Beatles. Photocathodes, phosphors, and 50,000 volts worked their magic on magnetically focused streams of electrons that amplified each incoming cosmic photon into a cascade of 100 million photoelectrically triggered photons. As these hit the final photocathode in a splotch, pattern recognition electronics linked to a computer—the analogue of the retina plus brain—reconstructed the location of the cosmic progenitor in the sky. The whole process was done in real time and took nanoseconds. Noise, ion spots, and sky background were electronically rejected so efficiently that even very faint objects could be observed repeatedly over several weeks, and their signals then co-added. This postprocessing helped compensate for the low sensitivity of the system, which was about 20%. Although the image-tube chain measured only 2 feet in length by 8 inches across, it was backed up with a crate of electronics as tall as a room.

Hired by Greenstein in 1959 as a Research Fellow in astronomy, Wallace Sargent brought to Caltech a fresh PhD from the University of Manchester in England. In that rigorous institution, he had been taught to be certain that measurements were correct before beginning to interpret them. His motto was "Better to measure the damned spectra, make sure you believe everything, *then* interpret."[185] A theorist by training, he was expected to work on how supernova explosions interact with the interstellar medium. However, six weeks into his fellowship, Sargent visited Mount Wilson Observatory and was "much taken with observing because it seemed easier than theory."[186] He promptly switched to stellar spectroscopy and measuring chemical abundances. Soon after, the challenge of exploring the high-redshift universe via the absorption lines in quasar spectra caught his attention, and his interests expanded to include intergalactic gas and the evolution and history of metal enrichment in galaxies.

In May of 1973, a chance encounter between Boksenberg and Sargent in the corridor of Robinson Hall at Caltech sparked a long-term collaboration, when "Boksy" convinced Sargent that his new instrument could amplify the power of the 200-inch telescope a thousand times. Sargent immediately recognized the potential of a linear detector for spectroscopy of very faint objects such as quasars, and promised copious observing time. Palomar was one of the few observatories where chunks of observing time were available to carry out large multiyear surveys. The deal was sealed in less than five seconds, according to Boksenberg, and by October the IPCS was mounted at the coudé focus of the 200-inch telescope. For the next several years "The Flying Circus"— as Boksenberg, his crated instrument, and his large technical support team became collectively known—commuted regularly between London and Palomar.

Chapter 9

BOKSENBERG'S FLYING CIRCUS: The Image-tube Photon Counting System (IPCS), along with youthful technician John Fordham, who designed—and sometimes shored up—the hardware and electronics, and PhD student Keith Shortridge, who expertly programmed the computer and handled software emergencies, would be flown in from London in advance of observing runs at Palomar. Before attaching the IPCS to the telescope, the technical crew worked inside the 200-inch dome for three or four days assembling the apparatus—said to resemble a pile of electronic junk held together by duct tape and baling wire. Boksenberg, aka "Boksy," arrived with a grab bag of postdocs and graduate students in time for the run, and Sargent drove to the mountain from Pasadena.

To these fun-loving Brits, Palomar and its local astronomers appeared to be rather stuffy and serious, and they lightheartedly refused to obey what they considered stilted codes of mountain conduct. Their antics emulated the irreverent characters of the popular 1969 British comedy series *Monte Python's Flying Circus*. So Boksy's team dubbed themselves "Boksenberg's Flying Circus." They were a colorful group to watch—wrecking at least one rental car and killing at least one deer on the mountain. Their antics sparked an observatory-wide ban on alcohol that was rescinded by Sargent as soon as he became director of Palomar in 1997.

Preparing for a typical night of observing, the team checked the focus of the telescope and calibrated their instrument, decided the program for the night (selecting targets, wavelength ranges, and gratings), snapped Polaroids of POSS prints to use as finding charts, and then grabbed a quick dinner at the Monastery. Back in the dome, observing was straightforward as they lingered on each object for long periods of time to maximize the signal-to-noise ratio of its recorded spectrum. Their digital observing system being interactive, the team crowded around a small oscilloscope in the coudé room to watch the spectrum build up photon by photon. The rate at which photons arrived might be as fast as several per second and spectral channel, or as slow as one per minute and spectral channel for the faintest objects in the sky. After about 5 minutes, a tiny-amplitude spectrum became visible on the screen. After a longer period, an absorption line might appear, too weak yet to do physics with, but enough to confirm that they had the correct object. The IPCS data were unsurpassed at the time, and the Flying Circus contributed substantially to Palomar's leadership in the burgeoning field of quasar absorption-line spectroscopy.

Quasar Absorption Lines: A Biometric Passport

Throughout the 1970s and 1980s, debates continued concerning the origin and evolution of quasar absorption lines, as the IPCS recorded tens of thousands of absorption lines in scores of quasar spectra. To disentangle and identify the dense jungle of lines, John Bahcall of Princeton University developed computer algorithms that tested the observed lines against tables of atomic transition probabilities and "cosmic" elemental abundances to find a best match.

Sargent, Peter Young, Boksenberg, and David Tytler, a University College London graduate student, subjected their bounty of relatively high-resolution quasar absorption line spectra to a battery of sophisticated statistical tests, and published their findings in a landmark paper in 1980.[187] First they found strong evidence that quasars themselves were not the source of the line-forming material, since there was no pileup of absorption lines around their redshift. Instead, they suggested that two physically distinct—and decidedly cosmological—absorbing "entities" were involved. Entities that formed multiple absorption line systems were dense, enriched with heavy metals, and located in the halos of intervening galaxies. Entities that produced single, sharp Lyman-α absorption lines were low-density intergalactic gas clouds of nearly pristine hydrogen and helium that filled the universe. These were the residua of galaxy formation.

To the present day, the hypothesized gas clouds responsible for the absorption lines have not been observed directly. Nor have satellite observations verified the existence of a hot intergalactic medium that would confine such clouds. Although Boksenberg's Flying Circus had achieved significant breakthroughs with the IPCS by boring a path to the young, otherwise-invisible universe, it would take larger telescopes with CCD detectors to gain further insights from quasar absorption lines. As spectral resolution and sensitivity increased, the number of observed absorption lines grew until they resembled a veritable dense forest of Lyman-α lines, sparsely punctuated with absorption lines from other elements that reflect the chemical composition of the known universe along the path to the quasar.

Analogous to growth rings in a tree, an increasing number of absorption lines in the spectra of quasars indicates an increasing distance to the quasars. Comparing two quasars, one at low and the other at high redshift (top and bottom graphs, respectively), illustrates the increasing number of intervening structures with increasing distance to the quasar. In these two spectrograms, quasars form an ideal backdrop for viewing absorption lines, as their nearly flat, featureless continuum extends far into the ultraviolet region of the electromagnetic spectrum. The broad Lyman-α emission line (the spike in the center)—a homology of quasar spectra—signals that a gaseous accretion disk or a collection of clouds surrounds the quasar's supermassive black hole. The narrow absorption lines that sharply pierce the continuum shortward (to the left) of Lyman-α are formed by intervening absorbing gas. At longer wavelengths, the few absorption lines are due to denser metal-rich structures such as gas disks in galaxies. Image assembled by William Keel from Keck data taken by Donna Womble.

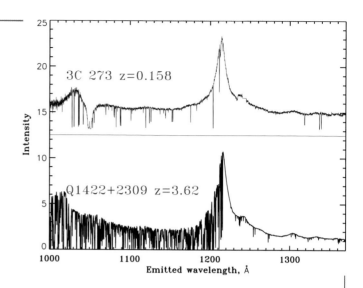

Quasars provided the means to explore, little by little, the high-redshift universe, where otherwise-invisible material was still collapsing into galaxies and stars. One important result was that at these high redshifts there appeared to be no deficit of baryons, unlike in the local universe. The reason is now understood to be the lower temperature (about 10,000 K to 20,000 K) of this very distant intergalactic gas, which greatly facilitates the detection of its baryons through absorption lines. This web of hydrogen is a cool precursor of the later web of hot gas at low redshifts.

Over cosmic time, an increasing fraction of intergalactic gas has since collapsed into denser and denser clouds and formed galaxies. The gas has been heated to temperatures of 100,000 K to 10,000,000 K by internal shocks and galactic outflows resulting from star formation and central-black-hole activity. These high temperatures have made it much harder to detect this warm-hot intergalactic medium (WHIM) in the local universe, leading to the notion of "missing" baryons. At this low redshift, the gas becomes invisible to ground-based telescopes and needs to be observed in X-rays.[188]

Despite being useful as distant background light sources for absorption line studies of the high-redshift intergalactic medium, quasars have one serious drawback: they emit a vast quantity of ionizing radiation. This affects the gas of their host galaxies severely. Thus, their spectra tell astronomers relatively little about the interstellar gas in the quasar's host galaxy itself. As if to redress this flaw, a high-redshift probe even more powerful than quasars—though capricious and short-lived—entered the arena in the waning 20th century.

Gamma-Ray Bursts: High-Beam Dazzlers

In the late 1960s, the Vela system of satellites monitored Earth and space for gamma rays that would signal surreptitious nuclear explosions in foreign countries. However, Vela instead found evidence for explosions of another kind: mysterious random gamma-ray flashes from space that appeared without warning, then rapidly faded. Thousands of these flashes, soon dubbed gamma-ray bursts or GRBs, were recorded by a succession of satellites while scientists debated their origin. Were the sources of these bursts relatively local, or in the halo of the Milky Way, or even extragalactic?

Gamma rays are energetic enough to travel unimpeded throughout the universe. They are the highest-energy, shortest-wavelength form of electromagnetic radiation. Unlike optical radiation, they are difficult to bring to a sharp focus. Therefore, during the pioneering days of gamma-ray astronomy, the positional accuracy of GRBs was crude, leaving astronomers to sort through thousands of faint stars and galaxies to find any potential optical counterparts. The Italian-Dutch satellite BeppoSAX, launched in April 1996, improved the odds by carrying an X-ray camera that rapidly pinpointed a burst's position in the sky and immediately transmitted it to ground-based observatories. The finding that many GRBs are followed by "afterglows" that spill over to progressively longer wavelengths—from the X-ray to the optical, infrared, and radio domains—effectively prolonged their detectability and brought hope that an optical spectrum could be obtained.

On May 8, 1997, Caltech astronomy professor George Djorgovski was observing with his students at the 200-inch telescope when radio astronomer Dale Frail[189] and BeppoSAX team member Marco Ferocci phoned him to announce a new gamma-ray burst. It was designated GRB 970508 after the date of its discovery. Time was of the essence: in contrast to the longevity of quasars, a gamma-ray burst's afterglow is unexpected and fleeting, like quicksilver. The window of opportunity for getting an optical spectrum of a GRB in order to study the composition of the gas in its host galaxy is essentially half a day after the burst happens. If astronomers don't acquire its spectrum in time, they just have to wait for the next GRB. Thus, to the delight of his students, Djorgovski immediately interrupted his own observing program at the prime focus and began taking a sequence of 5-minute CCD images covering the 10-arcminute error circle of the BeppoSAX observation. It was less than 6 hours after detection of the initial brief gamma-ray burst, and—unbeknownst to the astronomers then—the optical burst was growing brighter.

On the same night, the nearby 60-inch telescope took images of the surrounding field, including nearby photometric standard stars needed to calibrate data from the 200-inch telescope with that of the digitized Palomar Observatory Sky Survey (DPOSS). After two nights of observing, Djorgovski and colleague Mark Metzger had enough data to search for the burst's optical counterpart. Comparing their images to POSS, they found a variable object inside the error circle.

The source was beginning to fade as Djorgovski worked to decipher its light curve, and Metzger swiftly transmitted its coordinates to the Keck Observatory on Hawaii's Maunakea summit. On May 10, early on the first evening of his observing run on the Keck II 10-meter telescope, Chuck Steidel delayed his own observing program to take three 10-minute spectra of the fading source. He caught the burst's afterglow just before it sank below the western horizon. Metzger immediately analyzed the spectra sent by Steidel, and Djorgovski recognized some features that resembled the multiple absorption lines of intergalactic clouds in quasar spectra. A strong absorption line system at redshift 0.8 set a lower—but still clearly extragalactic—limit to the burst's distance. The upper limit to the burst's redshift was 2.01, since at higher redshifts Lyman-α would have been visible, and it wasn't.

A group of competing astronomers had raced to be first to identify the optical counterpart. However, by the time they phoned Keck Observatory to request a spectrum, they were told: "Too late!" After a couple of days of furious nonstop activity, Djorgovski and his collaborators announced the first credible identification and redshift of a gamma-ray burst afterglow, settling the decades-long debate over whether the bursts were local or extragalactic.[190] Whatever their genesis, gamma-ray bursts are the most powerful explosions in the known universe. Within seconds they release more energy than the sun will generate in its entire lifetime of 10 billion years. A gamma-ray burst's mercurial light dazzles us with 100,000 times the luminosity of its host galaxy, which itself may contain 100 to 200 billion stars. The objects producing these spectacular explosions may be extraordinarily intense supernovae, massive stars that implode to a black hole, or perhaps even more exotic phenomena.

Astronomers currently think that optical afterglows are synchrotron emission from electrons spiraling in shock-driven magnetic fields. Afterglow light curves, based in part on Palomar data, provided some first evidence that gamma-ray bursts are collimated into opposing, narrowly focused jets that are launched immediately after the explosion. Moving at close to the speed of light, these jets punch through the star's envelope within seconds to cast the innards of the star into space.

Recently, gamma-ray bursts have moved from being events of curiosity—and phenomenal physics—to being tools for analyzing the interplay between star formation and the interstellar medium in high-redshift galaxies. Unlike quasars, gamma-ray bursts do not immediately alter the local environment of their galaxy host. Through opportunistic spectroscopy of afterglows, astronomers can observe the process of star formation and metal enrichment at redshifts normally inaccessible for detailed studies of galaxy contents. Since gamma-ray bursts occur during the death throes of massive stars embedded in galaxy disks, they bestow valuable glimpses of the physical environments in star-forming regions where metals are produced. So luminous as to be detectable out to a redshift of 20 (!)—corresponding to a mere 180 million years after the big bang—they may one day uncloak the earliest generations of stars. Gamma-ray bursts may even elucidate the role they and quasars play in the reionization of the universe.

By turning their view from luminous objects to absorbing material, astronomers plunged much deeper into the early universe. Former evolutionary scenarios fell to cosmologists' new models of underlying dark matter pulling gas and galaxies along as it collapses into vast sheets, filamentary weblike structures, and knots. Since intergalactic gas is very tenuous, detecting its faint glow is at the very limit of current instrumentation. But detecting it is imperative: much of the universe's history is recorded in the cosmic web's complex and rich fabric. Strewn with ancient amniotic gas, the cosmic web serves both as a mother lode for baryons that accrete, cool, and condense into stars and galaxies, and as a receptacle for radiation and metal-enriched detritus spewed out from these same stars and galaxies.

Inside the bell jar. Photo by the author.

· 10 ·
From Ghosts to Galaxies: The Emergence of Structure

Chapter 10

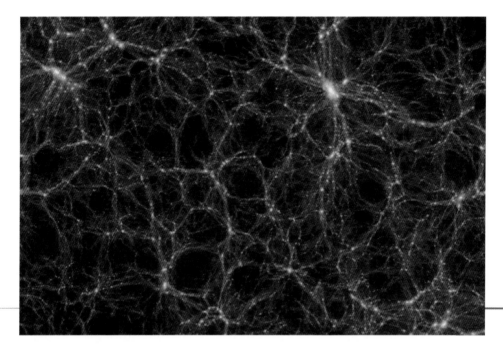

Astronomers and cosmologists have begun unraveling the early history of the universe, from its inception 13.7 billion years ago as an infinite-density singularity, to a hot, nearly featureless ball of expanding plasma, to its present state of large-scale structure and extraordinarily diverse phenomena. The observed spatial distribution of galaxies and features of the expansion imply that the universe is made of three components: normal matter (baryons), dark matter, and dark energy. The baryons are embedded in an intricate web of dark matter that has organized itself into intersecting filaments and sheets that surround voids spanning up to half a billion light-years. As the dark matter collapses under its own gravity, the baryons are pulled along. On smaller scales, the baryons collapse under their own self-gravity into gas clouds that form stars. Unlike dark matter, these stars emit light and are directly observable. Extensive computer simulations, such as the Millennium Run, shown here, and the Illustris Project, model these processes based on our current understanding of the nuts and bolts of the workings of the universe. These include the expansion of the universe, gravity, gas hydrodynamics, chemistry, radiation, magnetic fields, galactic outflows, star formation, and black holes. The goal is to model how synthetic galaxies would appear in the sky based on feedback from observational astronomers. The hope is that visible galaxies, as tracers of the large-scale structure of dark matter, will one day allow astronomers to plumb the depths of time and space as far as the big bang. The simulation shown here represents a slice—just under a billion light-years across—through the large-scale dark matter distribution in the present-day universe (at redshift zero). The yellow areas represent dense material forming galaxies and clusters of galaxies. Credit: V. Springel et al., "Simulations of the Formation, Evolution and Clustering of Galaxies and Quasars," *Nature* 435 (2005): 629, Millennium Simulation.

In the final paragraph of his book *The Realm of the Nebulae* published in 1936, Edwin Hubble wrote about the handicap astronomers face as they strain their vision at the dim boundary of the universe. There they measure shadows, then search among "ghostly errors of measurement for landmarks that are scarcely more substantial."[191] Yet, based on their searching of legions of such shadows, ghostly errors, and ethereal landmarks, by the end of the 1960s astronomers thought they had an inkling of how galaxies formed and evolved. Olin Eggen, Donald Lynden-Bell, and Allan Sandage had concluded from the motions of stars in the solar neighborhood that the Milky Way galaxy formed very early, during the rapid collapse of an immense gas cloud. The implication was that the morphological type—and fate—of a galaxy might be determined by its initial collapse history and would remain the same ever after. Only the galaxy's stellar content would slowly change with time, as new stars formed and short-lived massive stars exploded. Strengthening this notion, theoreticians argued that the arms of spiral galaxies may be long-lived density waves in the galaxies' disks that resist being wound up despite the disks' differential rotation. The lives of galaxies seemed serene and near-eternal.

Distant Shadows Red and Blue

The discoveries of quasars and of an explosive gaseous wind in the galaxy Messier 82 early in the 1960s suggested phases of possibly more recent and rapid evolution. The 1970s brought more signs of ongoing galaxy evolution as theoretician Beatrice Tinsley of Yale University modeled how stellar populations in galaxies age and transition from bright and blue to faint and red. Alar and Juri Toomre successfully explained some of Chip Arp's peculiar galaxies as the product of brief, but fierce, gravitational interactions of spiral galaxies. The Toomre brothers' models even suggested that such interactions might lead to orbital decay, mergers of galaxies, and the transformation of disk galaxies into ellipticals. Yet some astronomers seemed unprepared and even surprised when in 1978 Harvey Butcher and Augustus Oemler reported a discovery based on observations made at Kitt Peak National Observatory. Butcher and Oemler claimed that there seemed to be three times more blue galaxies in two distant rich clusters of galaxies than there were in nearby clusters. Could these distant blue galaxies mark distinctly younger galaxy populations at redshifts of 0.2 to 0.4 that had then aged significantly over the next 3 to 5 billion years, and largely disappeared from present-day clusters in our local universe? The inference that galaxy evolution had proceeded unexpectedly rapidly in such a short time was greeted with skepticism and brisk debate.

The normal route to understanding galaxy evolution had been to search for clues embedded in the structure and dynamics of nearby, present-epoch galaxies. Relying on data from distant galaxies at the technical limit of observability was a break from the norm. Since such galaxies appear as faint structureless blobs, Butcher and Oemler's claims of cluster membership were based solely on each galaxy's color and emplacement in the cluster. But might the blue galaxies be superposed foreground or background field galaxies (lone galaxies not part of a cluster)? Color measurement being a blunt tool, it was clear that detailed spectroscopic observations of the blue galaxies were needed to understand the galaxies' true nature. However, spectroscopy of such faint smudges was beyond the reach of even the new 4-meter telescope at Kitt Peak. The culprit: photographic plates that recorded incoming photons with an efficiency of only about 1%.

Just at that time, instrumentation at Palomar was making giant strides. The Jet Propulsion Laboratory was leveraging state-of-the-art instrumentation in the form of charge-coupled devices (CCDs), adaptive optics, and more to keep Palomar at the forefront of astrophysical research. Because of their close links with JPL, Jim Gunn and Jim Westphal were privileged to experiment with such "space age" technology as CCDs before they became generally available. These photoelectric detectors are 20–80 times more efficient than photographic plates in recording photons. Using components he cobbled together from various sources—a used 135 mm f/2 Xero Nikkor lens, a new now-classic and unobtainable 58 mm f/1.2 NoctNikkor lens, and lots of small optical components from the iconic Edmund Scientific catalog with which he had been dealing since high school in the late 1950s—Gunn designed and fabricated a versatile instrument specifically for the new CCD detector technology. Small by today's standards—no bigger than a suitcase—the instrument nestled into the prime-focus cage of the 200-inch telescope, where a photographic camera or spectrograph would normally sit. The instrument could take both direct images and spectra, so Gunn named it "Prime-Focus Universal Extragalactic Instrument,"[192] with the acronym PFUEI (pronounced, facetiously, "phooey"). Its detector was a CCD the size of a postage stamp, yet it was nearly two orders of magnitude more efficient than a photographic plate. In minutes, it captured images of faint galaxies that had previously taken hours, and in just a few hours it took spectra of galaxies that had previously been unobtainable. As PFUEI was also a gateway instrument to the 4-Shooter, itself a prototype for the Wide Field and Planetary Camera of the Hubble Space Telescope, Gunn couldn't wait to take PFUEI to the 200-inch telescope.

In 1980, Gunn joined with Carnegie astronomer Alan Dressler to study galaxy evolution in seven distant clusters. With the new combined CCD camera and spectrograph, they hoped to reach redshifts of 0.35 to 0.55, then considered the very limit of feasibility for the 200-inch telescope. Among the objects on their observing list were two familiar clusters, one containing the radio galaxy 3C 295 identified by Rudolph Minkowski in 1960, and the other cluster Cl 0024+1654 discovered by Milton Humason in 1956.

Soliloquies from the Cage: "I was strongly motivated to observe by two things: It was nearly always the science I was working on, and I mostly observed with instruments of my own devising and manufacture. So just babying a device and coddling it along was not a negligible challenge. When I wasn't busy with an instrument, I looked up at the sky. I don't know whether there are words to describe my experiences at the prime focus—it was beautiful beyond any kind of words. It was my very favorite place on Earth. It's being very much alone. It can be very cold. It could be either incredibly hectic or incredibly peaceful, depending on whether the particular instrument I was dealing with was working or not. When I first started observing at the prime focus, I was alone with a night assistant—Gary Tuton or Juan Carrasco—who operated the telescope. There were no fancy electronic gizmos, just photographic plates. Nothing can go wrong with the photographic plate, at least not that you know about while you're exposing it. It was an extremely peaceful experience. There was no automatic guider, so I manually kept the guide star on the crosshairs. There was an incredibly poor-quality sound system in the prime focus, but it was one of the few times I could listen to an entire Wagner opera in one sitting.

"Later, with complex electronic instruments such as the SIT spectrograph, the double spectrograph, then PFUEI and 4-Shooter, all requiring a fair amount of computer support, observing became a communal effort. I was joined by Bev Oke and Barbara Zimmerman—an expert in coding for the electronic instruments. Either Bev was extremely nice and let me ride at prime focus all the time, or he didn't like it very much. Anyway, I was always the guy at prime focus, which suited me just fine. I was 100 some-odd feet above the ground, and there was nobody else in the world. It was just me and the sky. I miss it, I really do.

"I feel that astronomy is what I should be doing and was put here to do. I have loved it since I was very young—age 7—and read my first astronomy book. Being hooked on amateur astronomy, I persuaded my dad to help me build a 4-inch reflector and several small refractors. I read everything I could get my paws on about the subject, and there was no question about what I was going to do with my life."

(J. E. Gunn, personal interviews with the author, 2008–2019)

Chapter 10

In a paper published in 1982, Dressler and Gunn reported some first results on these two clusters. During four nights of hard work with PFUEI at the 200-inch telescope, they had managed to record spectra of 17 faint smudges—among the brightest candidates in each cluster field—and to measure redshifts for 13 of them. The results were surprising. Whereas in Cl 0024+1654 all six galaxies were at the same redshift and, therefore, likely cluster members, in the 3C 295 cluster only five of the seven measurable galaxies were cluster members. The other two were lower-redshift foreground galaxies. Such foreground contamination would make it more difficult to learn about the cluster's true properties. But most surprising were the spectra of the blue galaxy candidates pointed out by Butcher and Oemler. Of seven blue objects that were candidate cluster members, only one had emission lines typical of a normal spiral galaxy. Two others showed emission lines typical of Seyfert 2 galaxies, which are spirals with an active nucleus. The spectra of the other four blue galaxies had no discernible features, making it clear that spectra with higher signal-to-noise ratios would be needed. Clearly, the blue galaxies pointed out by Butcher and Oemler were candidate members of the two distant clusters. Yet they were not spectroscopically normal spirals of the kind seen in nearby clusters. So what were they?

E+A Galaxies

To address this question, the quality of the spectra would have to be improved. This meant that at least 20 to 30 galaxies in each cluster would need to be observed, each with double or even quadruple the exposure times. Yet, having taken turns observing with PFUEI in the prime-focus cage for four nights, the astronomers knew their goal would be difficult to achieve. One night, having placed the single slit of the PFUEI spectrograph across two galaxies orbiting within a distant cluster, Dressler sat in the prime-focus cage patiently guiding the telescope. While the light went down the slit for the two target galaxies, light from nearby galaxies was reflected by the slit jaws back into the night sky, its precious information lost forever. If only one could take spectra of 5 to 10 galaxies at a time, he mused. He had learned that Kitt Peak National Observatory was in the process of trying to replace single slits with multiple *slitlets* to capture light from several galaxies simultaneously. Feeling that the primacy of the world's largest telescope was being challenged, Dressler and Gunn borrowed the concept of multiplexing apertures and replaced the single slit of PFUEI with a dozen customized photo-negative masks. The masks channeled the light of up to a dozen galaxies simultaneously through small, precisely configured slitlike "apertures," which were transparent rectangles in the film. Now called "multislit spectroscopy," this technique has become a staple of extragalactic astronomy.

By 1983, Dressler and Gunn had collected 20 new spectra of galaxies in the 3C 295 cluster in only three nights at the 200-inch telescope. Combined with their earlier observations, they now had new spectra of 26 galaxies, of which only 12 turned out to be cluster members; the remainder were foreground or background contaminants. The spectra of the 12 members were perplexing. The reddish cluster elliptical and S0 (lenticular) galaxies were similar to those of local cluster galaxies, and thus showed no significant evolutionary changes over the past 5 billion years.[193] Some of the blue galaxies had spectral features that Dressler and Gunn hadn't seen before in cluster galaxies. It was a struggle to identify the faint, fuzzy lines of some of the blue galaxies until their spectrograms were co-added. What they saw appeared contradictory: broad Balmer absorption lines, characteristic of young type-A main-sequence stars, appeared alongside the H + K lines of ionized calcium, typical of late-type giant stars in old populations. Oddly, the three blue cluster galaxies showed clear signs of having two distinct stellar populations. These strange objects resembled old elliptical or S0 galaxies at high redshift—yet they also featured young post-starburst populations similar to those seen in some nearby dwarf galaxies.

Balmer lines can be tricky to interpret. They can be the signature of robust star formation that began about 100 million to 1 billion years before the light left the galaxy and then rapidly ended. Was, then, this odd pairing of young and old stellar populations a sign that an episode of intense star formation had been quenched in these distant cluster galaxies? Seemingly, these blue "E+A" galaxies—as they were soon called—were significantly more common in the past. The E+A designation encapsulates the resemblance of the spectra to old elliptical galaxies with significant light from young A-type stars added in. It also makes the point that the starburst phenomenon occurs on galactic scales.

> "Akin to a religious tenet, there was a strong commitment to the Cosmological Principle—that we do not live in a special place nor in a special time. So to have to say, 'We just don't see those things today, but they were around 3 or 4 billion years ago in fair abundance,' was very uncomfortable. I think it's taken all these years, and being able to carefully model with computers and simulations how the universe is evolving, to get to this idea that we do live in a special time: the end of star formation."
>
> (A. Dressler, personal interviews with the author, 2010–2014)

Among the six blue cluster members, three were E+A galaxies and three were blue Seyfert-type galaxies. That star-forming galaxies with active nuclei were present—and quite numerous—in a rich galaxy cluster five billion years ago was unexpected. It was splendid spectroscopic confirmation of Butcher and Oemler's guess, five years earlier, that the increased numbers of blue galaxies in clusters at redshifts of about 0.4 to 0.5 might be a direct sign of galaxy evolution between then and now. The 200-inch telescope had again yielded crucial direct evidence of galaxy evolution over the most recent third of the history of the universe.

Ring of Fire

What process within the rich clusters produced these intriguing E+A galaxies? Dressler and Gunn noticed that galaxies in the outskirts of their seven clusters seemed to form stars more actively than those closer to the centers of the clusters, which had ceased forming stars. Why are local clusters filled with galaxies whose gas is gone? In 1951 Walter Baade and Lyman Spitzer[194] had suggested that spiral galaxies in dense clusters get stripped of their interstellar gas by frequent interactions either with other cluster galaxies or with a hot intracluster medium. Direct collisions between galaxies would sweep their interstellar material out into intergalactic space, preventing star formation in cluster spirals. Two decades later, Gunn and Richard Gott[195] published a theoretical paper outlining how galaxy clusters grow in mass as gas and galaxies fall into them at high speeds. The clusters become filled with gas heated by shocks to 10,000,000 to 100,000,000 K. Unable to cool over the age of the universe because of its low density, this hot gas just sits there. Meanwhile, galaxies fall in from filaments that make up the large-scale structure of the universe. Like a driver in a convertible, the galaxies moving through this hot gas feel a wind as their native interstellar gas is being stripped away by the cluster gas, a mechanism called "ram-pressure stripping." During the ride inward, some fraction of a galaxy's gas is compressed and jolted into brisk star formation, while the remainder is stripped away and disperses into the intracluster medium.

From Ghosts to Galaxies: The Emergence of Structure

A galaxy's physical environment can play a fundamental role in its evolution. The spiral galaxy ESO 137-001 is being stripped of its gas as it plows through the 100-million-degree intergalactic medium of the Norma Cluster. While the galaxy's native stars remain gravitationally bound, new starbirth has migrated to its long train of stripped gas. After considering various theories—picturesquely named galaxy harassment, quenching, and starvation—to explain why some cluster galaxies cease forming stars, many astronomers seem to have settled on ram-pressure stripping. Credit: Composite optical and X-ray image; X-ray: NASA/CXC/UAH/M. Sun et al.; optical: NASA, ESA, & the Hubble Heritage Team (STScI/AURA).

To explain their new observations, Dressler and Gunn proposed that the apparent stratification in the rate of star formation occurs over the roughly one billion years that galaxies take to reach the central region. During this time, their A-stars evolve and fade until the E+A spectrum is gone. This hypothesis seemed to hang together, with the caveat that it was based on a mere snapshot in time. The fact that there are blue galaxies in the outskirts of the cluster does not in itself argue for ram-pressure stripping. Mergers between galaxies, which in clusters tend to occur preferentially in the outskirts, could be culpable as well. As Gunn recently mused, "The facts were undisputed. How to interpret them was not very clear."[196]

Chapter 10

Deciphering the distorted shapes of distant cluster galaxies and deciding whether starbursts were due to mergers or the stripping of gas awaited the launch in 1990 of the Hubble Space Telescope with its high-resolution cameras. Before then, few galaxies were optically accessible beyond a redshift of 0.5, and little was directly known about run-of-the-mill galaxies beyond a redshift of a few tenths—considered a high redshift before the launch. It was rare to find special objects, such as radio sources or other active galaxies, with redshifts larger than 1—so rare that entire papers were devoted to them. In those days, some astronomers spent most of their careers searching for and studying a handful of extremely distant objects.

With the higher resolution of the Hubble Space Telescope, astronomers identified the blue smudges as spiral galaxies. The spirals—appearing distorted and ragged—were more plentiful in distant clusters, appearing mainly in the clusters' outer regions. Since then, evidence that star formation is enhanced in galaxies falling into clusters has continued to grow. At present, there is ample evidence out to redshifts of 7 and higher that galaxies evolve. In many nearby, present-day clusters, massive galaxies no longer form stars—that privilege has become the domain of local field and dwarf galaxies. Yet this cannot have been true forever and must have changed in a cosmic epoch that could not be reached by observational astronomers of the 1980s.

The next jump in access to the high-redshift universe came not from improved technology, but from the clever exploitation of existing techniques that had been in place at Palomar for decades.

The Far Side of the Universe

In 1984, when Chuck Steidel began graduate studies at Caltech, quasar absorption lines provided the sole means of detecting intervening gas clouds and galaxies at high redshifts. With a generous allotment of observing time on the 200-inch telescope, he and his advisor Wal Sargent studied distant galaxies in an oblique way: not by their direct emission, but by the spectral absorption lines they imprint on the light of bright background quasars. Metals in the galaxies' interstellar gas, such as C IV (triply ionized carbon, produced only by stars), helped them trace the metal content of galaxies over cosmic time. Lyman-limit absorption systems, which are clouds of neutral hydrogen in the intergalactic medium dense enough to block out a quasar's ultraviolet light, helped them trace the neutral hydrogen. Combined, these absorption features in quasar spectra provided a first glimpse of the chemical enrichment history in intervening galaxies.[197]

However, it was frustrating to Steidel that these galaxies were forming stars and producing heavy elements at redshifts higher than astronomers could observe in a direct, systematic way. The techniques and instruments of that era were simply not up to the task.

On one hand, being a spectroscopist trained by Sargent to look down his nose at mere pictures, Steidel knew that there had to be some connection between this universe of galaxies and the background quasars known from the survey by Gunn, Schmidt, and Schneider. On the other hand, the surveys also made him realize that no one believes anything unless you can point to an image. By the time he left Caltech for a postdoctoral fellowship at Berkeley, Steidel's roots in the subject of quasar absorption lines had grown deep. Now rubbing shoulders with people interested not just in quasar absorption lines but in galaxies at large, he decided that, rather than continuing to work exclusively on absorption line systems, his next step would be to search directly for the entities— perhaps scraps of young galaxies—that create the absorption lines.

Steidel had followed closely the pioneering work of Jacqueline Bergeron, a European astronomer who used deep images and spectra of spatially resolved objects adjacent to quasars to identify the sources responsible for producing the absorption line systems in the background quasar spectra. In so doing, she found that these absorption line systems were caused by the extended gaseous halos of galaxies in or near the light path.[198] However, her work during the late 1980s remained restricted to nearby objects, and Steidel wanted to push further out.

There are thousands of galaxies visible on a typical deep CCD image taken at the prime focus of the 200-inch telescope. But which ones are far away? There is no depth perception in a two-dimensional image of the three-dimensional sky. Objects that are faint, small, or red might be nearby, in the vicinity of the Milky Way, or very far away, in the most distant observable reaches of the universe. Since the faintest galaxies do not necessarily correspond to the most distant ones, it would normally require inefficient, time-consuming spectroscopy to find the ones that are distant.

The concept of inferring distances to galaxies by comparing their brightness in various broadband color filters had been around for decades. The general technique was introduced in a 1957 publication by Carnegie astrophysicist Bill Baum, who described the shifting of a galaxy's spectral energy distribution through various band passes, defined by color filters, as its redshift increases.[199] By 1962, Baum was able to show an international audience of extragalactic astronomers how he determined the distances to three clusters of galaxies at redshifts of 0.19, 0.29, and 0.44 from photoelectric photometry in 9 filters with the 200-inch telescope.[200] At that time, Baum knew of only one cluster of galaxies about a quarter as far away as Steidel was now aiming for, and he could hardly have imagined such a drastic spectral signature as the Lyman-limit depression. Others soon followed Baum: Allan Sandage picked out radio-quiet quasars based on their peculiar colors in 1965; graduate student Richard Green used broadband colors to select quasar candidates for follow-up spectroscopy in 1976; David Meier proposed that images in three filters were sufficient to identify high-redshift galaxies in 1976; and Maarten Schmidt, Jim Gunn, and Don Schneider developed a spectroscopic method to hunt for high-redshift quasars in 1984—but didn't push it to the level required to see the much fainter *galaxies* at similarly high redshifts.

Sargent's large survey for Lyman-limit absorption systems had triggered some new insights. The ultraviolet light of continuously star-forming galaxies is dominated by generations of young massive stars, which produce copious amounts of ionizing ultraviolet radiation. Thus, the intrinsic shape of a galaxy's spectral continuum is approximately flat. However, the spectral continuum we see from Earth is not nearly flat. Some ultraviolet radiation is absorbed by hydrogen clouds within the host galaxy. As the remaining light propagates toward Earth, further ultraviolet radiation is absorbed by neutral hydrogen in the cosmic web. The farther the light travels through clouds of intergalactic hydrogen, the more the ultraviolet component of the light is absorbed and removed from a distant galaxy's spectrum. As a result, the intrinsic spectrum of a distant star-forming galaxy is severely attenuated shortward of the ionization energy of hydrogen. This drop in the observed spectral energy distribution is called the "Lyman break" at $912 \times (1 + z)$ angstroms, where z is the galaxy redshift. The Lyman break is a defining spectral signature of star-forming galaxies not mimicked by any other objects.

However, there is a catch: The Earth's atmosphere is opaque to cosmic radiation shortward of 3,000 angstroms. So the Lyman break is observable with ground-based telescopes only in galaxies with redshifts larger than about 2.3. Galaxies at this redshift are of special interest because they are less than 3 billion years old and still intensely forming stars. Thus, extending the existing concept of multicolor techniques, Steidel began a search for star-forming galaxies at redshifts higher than 3 by devising a system of three filters designed to straddle the redshifted Lyman break. He enlisted the help of Don Hamilton, a Caltech research associate who had perfected similar broadband imaging, to learn the craft of deep imaging and to fabricate a custom ultraviolet filter. Considering the project to be high-risk, the two wrote three papers on their technique and how they expected things to look before a single image was obtained. Although the initial goal was to examine galaxies near the lines of sight to quasars, the Lyman-break technique would turn out to also be very useful for general field galaxies.

Having no access to Palomar at the time, Steidel and his colleagues tested their concept at a multitude of other observatories, using the UK's 4.2-meter William Herschel telescope at Roque de los Muchachos Observatory in the Canary Islands and the 4-meter Blanco telescope at Cerro Tololo Inter-American Observatory in Chile, among others. Steidel found the variance of these first observations, sometimes obtained in poor seeing and of modest depth and breadth, unsatisfactory. Fortunately, he gained access to Palomar again when he returned to Caltech from the Massachusetts Institute of Technology in 1995, this time as a professor. Within days of his return, Steidel "dragged" Kurt Adelberger, his first-year graduate student from MIT, and British astronomer Max Pettini to the prime focus of the 200-inch telescope, where they racked up images in three colors. The first field they observed—a strip of sky designated Small Selected Area 22 (SSA22) and covering two adjacent 9 × 9 square-arcminute fields—would turn out to yield major findings.

> "I collected a lot of random bits of data during these years—a 'dog's breakfast' of data from many different places and observatories and with different techniques—and it was really hard to put it together."
>
> (C. C. Steidel, personal interviews with the author, 2009–2011)

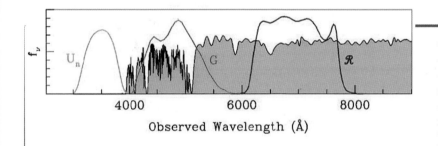

The Lyman-break technique enabled the wholesale identification of galaxies that formed early in the universe's history. Here the spectrum of a model star-forming galaxy (yellow) is shown with a redshift of 3.15 so that its Lyman limit appears in the optical at about 4,000 angstroms. The spectral sensitivities of the three custom filters, through which sky fields were imaged at the Palomar 200-inch telescope, are drawn in. The near-ultraviolet-band filter U (blue curve) lies on the short-wavelength side of the Lyman break for galaxies at redshift 3, the green filter G lies on the long-wavelength side, and the red filter R helps distinguish highly reddened or intrinsically red sources in the local universe. To search for the telltale drop in spectral intensity, deep images of galaxy fields were taken through these three filters. Galaxies that were visible in the green and red filters, but had dropped out of the ultraviolet filter, would likely lie at redshift 3 or slightly higher. In contrast, foreground sources would remain visible in the ultraviolet filter. (C. C. Steidel, M. Pettini, and D. Hamilton, "Lyman Imaging of High-Redshift Galaxies. III. New Observations of Four QSO Fields," *Astronomical Journal* 110 [1995]: 2519.) Figure 2a from C. C. Steidel, "Observing the Epoch of Galaxy Formation," *PNAS* 96, no. 8 (1999): 4232–4235, Copyright (1999) National Academy of Sciences, U.S.A.

Instead of requiring numerous time-consuming spectra in each field to weed out nearby objects, the Lyman-break technique required just three filtered exposures to identify with high efficiency the objects likely to be extremely distant galaxies. Observing was still labor-intensive, with ultraviolet exposures taking 10 hours of 200-inch telescope time, and exposures through the red and green filters taking two hours each. With some delays due to inclement weather, Steidel and colleagues spent four nights imaging each of nearly 20 fields. Follow-up spectroscopy at the Keck telescope on Maunakea in Hawaii confirmed whether dropout candidates were at high redshifts. This technique is essentially free of selection effects, with only minor contamination by low-redshift objects.

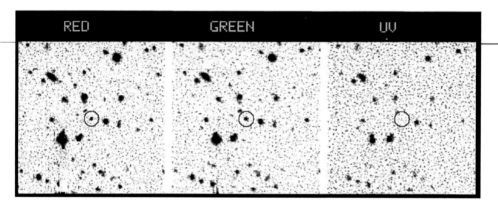

Small portions of three images taken at the 200-inch telescope to search for ultraviolet "dropouts." The circled galaxy at the center of each image is visible in the red and green filters but invisible in the ultraviolet filter, confirming that its redshift is around 3. The filters were designed to detect dropouts in a narrow redshift range of 2.6 to 3.4, which corresponds to a look-back time of around 11.5 billion light-years from Earth. Figure 2b from C. C. Steidel, "Observing the Epoch of Galaxy Formation," *PNAS* 96, no. 8 (1999): 4232–4235, Copyright (1999) National Academy of Sciences, U.S.A.

Steidel was genuinely surprised when follow-up spectroscopy at the Keck 10-meter telescope verified that most of the candidate galaxies that dropped out in the ultraviolet images had redshifts around 3 and weren't oddball local objects. The astronomical community—at first skeptical about any scheme to rapidly identify high-redshift galaxies—embraced and widely copied the technique. Whereas during the 1980s the upper redshift limit for galaxies seemed stuck at around 1, which corresponds to an epoch slightly more than halfway to the beginning of the universe, it suddenly nearly tripled, giving astronomers a glimpse of the more distant, younger universe. Instead of the rare discovery of a galaxy at high redshift, such as Minkowski's 3C 295 or a distant quasar, the newly christened "Lyman-break galaxies"[201] became a gateway to the formation and history of galaxies in the high-redshift universe. Soon catalogs of hundreds of these objects were available, enabling astronomers to systematically study star-forming galaxies and their large-scale distribution at only about 15% of the current age of the universe.[202] As soon as Steidel had gained access to these early galaxies and their stellar populations, metallicities, and star-forming properties, he dropped his work on quasar absorption line surveys "like a hot potato"[203] and focused on this promising new research field.

Galaxy Clustering and Lyman-α Blobs

One night, Steidel and his students were taking spectra of Lyman-break candidates that were located within the small strip of sky designated SSA22. As data accumulated in real time on the monitors at Keck, each spectrum seemed oddly similar to the one before. Finally, Steidel exclaimed, "Holy smokes! All of these galaxies are at the same redshift!"[204] His impression was verified when a histogram of the measured redshifts showed a conspicuous pileup of 16 galaxies around redshift 3.09.[205] Clearly, although for these galaxies only 2 billion years had elapsed since the big bang, they seemed to be already strongly clustered. For the first time, empirical data existed to support theoreticians' predictions of galaxy clustering early on in the universe.

The spike in the redshifts of these galaxies piqued the observers' curiosity. Might these galaxies represent only the most luminous among many more faint sources at that redshift? Steidel and his students began searching for fainter galaxies, capitalizing on two facts. First, star-forming galaxies emit prodigious amounts of Lyman-α as their hydrogen in the interstellar medium recombines. Second, a strong Lyman-α emission line is always easier than continuum radiation to spot against a noisy sky background. In the spring of 1997, Steidel ordered a custom narrowband filter to match the redshifted wavelengths of Lyman-α lines that might be emitted by any galaxies in that spike. Then, hour after hour, he took CCD exposures of SSA22 with the 200-inch telescope. The sources were so faint, compared to the night sky, foreground galaxies, and cosmic rays in the CCD frames, that his student Kurt Adelberger spent weeks afterward analyzing the data before he could extract any useful results. His time was well spent. Sources up to two magnitudes fainter than the original Lyman-break galaxies popped up in the data, 72 in all, each with excess

Lyman-α emission. The estimated mass of the entire "cluster" of galaxies corresponded to about 10^{15} solar masses, comparable to the masses of rich clusters in the nearby universe. Was this then a high-redshift snapshot of a region destined to become a rich cluster of galaxies?

Unexpectedly, there were two extensive Lyman-α-bright hazes surrounding some of the galaxies. The gas, dense and bright in the central parts, tapered off until it disappeared into the intergalactic medium. Based on the images, Steidel playfully named the two large glowing gas clouds "blob 1" and "blob 2" after the 1958 science fiction film *The Blob*, which depicts giant amoeba-like alien monsters with voracious appetites. The name stuck, and a new class of objects was born. Lyman-α blobs are spatially extended regions that emit mainly Lyman-α photons, but no continuum.[206] Blob 1 in SSA22 measures 150–200 kiloparsecs across and is several times larger than the entire Milky Way galaxy.

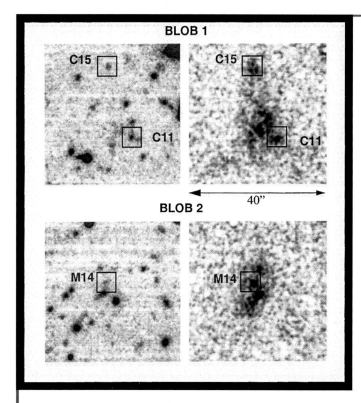

The two left panels show small portions of images from the discovery paper, published in 2000, that describes the original Lyman-break survey done at the 200-inch telescope. Blue and visual images have been co-added. The panels on the right show corresponding portions of deep Lyman-α follow-up images centered on blob 1 and blob 2. Lyman-break galaxies within the cluster at redshift 3.09 are marked by squares. Far larger than the galaxies they enshroud, blobs are among the largest known luminous objects in the early universe. Their soft glow has traveled 11 billion years to reach Earth. Blobs are now known to contain galaxies in their formation phase, some with active galactic nuclei and others with more mature stellar populations. Image from Figure 6 from C. C. Steidel et al., "Lyman-α Imaging of a Proto-cluster Region at <z> = 3.09," *Astrophysical Journal* 532 (2000): 170–182, © AAS. Reproduced with permission.

Still somewhat mysterious, these galaxy overdensities and Lyman-α blobs are not clusters as they appear today, since gravity has not yet had sufficient time to draw them close together. These "proto-cluster" regions were aggregates in the universe when it was only about 16% of its present age, so they reflect the distribution of matter in the young universe soon after the big bang. Perhaps we are seeing the progenitors of rich clusters of galaxies at a time when—under the influence of their own gravity—they had begun decoupling from the Hubble expansion.

Since the discovery of galaxy overdensities and Lyman-α blobs, continued imaging at the 200-inch telescope has identified density spikes in virtually every field surveyed, and scores more high-redshift proto-clusters and blobs have been discovered with telescopes across the world. Lyman-α blobs are now known to be numerous and a significant component of the young universe. Given that some of them appear to line up with the large-scale filamentary structure, they may provide clues to how galaxies form and how they relate to the growth of structure in the early universe.

Violent Winds

Steidel and Adelberger had expected to find evidence that gas from the intergalactic medium is falling into Lyman-break galaxies and feeding their star formation. Yet they found the opposite: powerful winds are hurtling *outward* from these galaxies at up to 2,000 km/s. Driven by supernova explosions and radiation pressure that inevitably accompany intense star formation, these superwinds may be strong enough to carry ordinary matter past the gravitational grasp of the galaxies' dominant dark matter.

Especially at high redshifts, superwinds may play an important role in galaxy evolution by sweeping hydrogen, helium, and heavy elements out of galaxies and depositing them into the vast intergalactic medium. In turn, gas that falls into galaxies, and hence the new generations of stars that form from this gas, are enriched by the fresh deposition. At the highest redshifts currently observed, the intergalactic medium shows signs of having already been a bit enriched with metals. What drives these superwinds? Merger-induced starbursts or active galactic nuclei? Does the ejected metal-enriched gas flow outside the galaxy to sojourn for a couple of billion years—getting spun up by whatever density fluctuations exist in the nearby environment—and then trickle back in a recycling program of cosmic proportions?

Galactic superwinds have some interesting consequences. Most elements were not created in the big bang, but in nuclear reactions deep in stars. How then did newly created elements travel from the centers of stars to our bodies? It was long thought that when stars

Chapter 10

explode as supernovae, their enriched detritus remains confined to the local environment. The discovery of raging superwinds means that enriched material from a foreign galaxy's starburst, occurring long before Earth formed, may have traveled through vast expanses of space to enrich our galactic neighborhood, and with it our solar system and—ultimately—our own bodies.

Galaxies consist of more than just their star-bright inner regions. Astronomers have discovered increasingly extended and complex outer structures not previously seen. In cosmic-web filaments, many galaxies are surrounded by relatively dense hydrogen gas. In a recycling galactic ecosystem, cold gas flows from the cosmic web to the inner parts of galaxies and feeds star formation there. In turn, hot inner gas is sporadically expelled by supernovae, massive stars, and occasionally an active galactic nucleus, thus enlarging the galaxies' presence.

Because of the way Lyman-α photons propagate and scatter, a star-forming galaxy generating a bounty of ionizing photons can appear enveloped in a Lyman-α cocoon that glows out to large radii. This cocoon can be extremely faint if the star formation rate is of order one solar mass per year, but can also be very bright—like a Lyman-α blob—if the star formation rate is of order 1,000 solar masses per year. Thus, the few observed Lyman-α blobs may represent only the tip of an iceberg of ordinary Lyman-α halos surrounding galaxies.

About half a century ago, astronomers first noticed the gaseous filaments of the cosmic web in the form of quasar absorption lines. By observing such lines in quasars distributed across the sky, astronomers sampled the metallicity, structure, kinematics, and evolution of the intergalactic medium across vast spaces. Their observations have served to vet theoretical models of the structure and evolution of the universe and its contents, including its baryons, dark matter, and dark energy. Yet at present data remain sparse, especially concerning direct images of the cosmic web from its intergalactic medium. Attempts have succeeded in imaging the relatively bright emission surrounding quasars and other strongly ionizing sources. Some especially creative teams of astronomers and instrument builders hope to image infalling gas streams that glow through collisional excitation as they enter galaxies from filaments of the cosmic web. While ultrasensitive instruments are at various stages of design, testing, and commissioning at Palomar and elsewhere, they are still near the limit of feasibility. The history of astronomical instrumentation assures us, however, that it is only a matter of time until the splendor of the cosmic web is fully revealed.

Counterbalancing supports, seen from below the 200-inch mirror cell. Photo by the author.

· 11 ·

Solar System Shuffle

Chapter 11

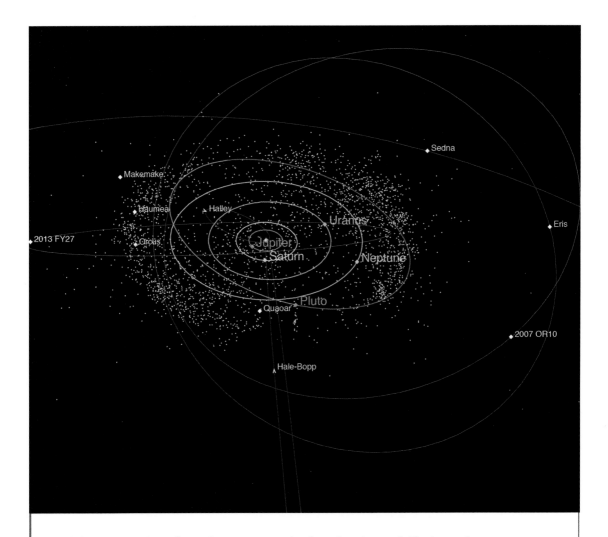

The pre-1990 view of our solar system was simple and uncluttered: The inner planets were rocky, the outer planets were gas giants, a rocky asteroid belt filled the void in between, and ice dwarfs orbited beyond Pluto. The view as of January 2018 is illustrated in this schematic oblique map of the outer solar system, seen from an angle 38 degrees above the ecliptic (the plane in which the major planets of the solar system orbit). Besides the main planets Jupiter, Saturn, Uranus, and Neptune there are vast numbers of smaller bodies (depicted in yellow) orbiting in the Kuiper belt, including the dwarf planet Pluto. The planets formed from material in the protoplanetary disk surrounding the young sun. Leftover debris then congregated in distinct bands of material called the asteroid belt (between Mars and Jupiter, not shown here), the Kuiper belt (shown here in yellow), and the Oort cloud (beyond the boundaries of this map). Credit: NASA/JPL-Caltech/P. Chodas.

"Astronomers deal with things far removed from human experience. Geologists are more involved in the real world,"[207] opined former Jet Propulsion Laboratory director Bruce Murray, who began his career in 1960 as a postdoctoral fellow in geology at Caltech. Fresh out of the Air Force, Murray had discovered "Space" and came to Pasadena with a spirit of adventure, aspiring to study planetary surfaces at Caltech. He quickly befriended Jim Westphal, a homespun, brilliant field seismologist with only a bachelor's degree in physics. Both men had worked in the oil fields, so they had a common set of experiences that were different from those of most people at Caltech. The self-proclaimed "buccaneers" were young and audacious, with an abundance of energy and few inhibitions. More importantly, Westphal was an instrument builder par excellence, and Murray still had connections to the Air Force: his commission and security clearance.

The Buccaneers

While transitioning from the Air Force to postdoctoral fellow, Murray needed access to some military detectors to observe very faint solar system objects. He boldly arranged a visit to the Naval Air Weapons Station at China Lake near California's Mojave Desert. While there, he "practiced glad-handing and talked up what I wanted to do at Palomar Observatory."[208] The upshot was that he walked out with eight scarce state-of-the-art infrared military detectors that his "newfound buddies" had stuffed into his pockets. In return, his buddies wanted Murray to tell them what he learned about the planets and about the properties of their detectors. With Westphal's help, the detectors were installed on the Palomar telescopes. Together the two men built a new photometer featuring quantum detectors and cryogenic liquid-nitrogen-cooled infrared detectors—a Westphal invention to beat down noise in the data. Their instruments were 50 times more sensitive than previous ones, giving Murray and Westphal an edge in the infrared domain that would usher in planetary observations at Palomar.

With special permission from the director of Palomar, infrared astronomers were allowed to squeeze in observing time before and after the main optical astronomer's assigned time. At first, they viewed relatively bright objects, such as the moon, Venus, Mars, and Jupiter, that didn't require dark nights. Murray and Westphal would begin their infrared observations at early dusk, one and a half hours before the main astronomer's dark time. They would then wait through the dark hours of the night until they were allowed to take over the telescope again at dawn for another one and a half hours. Westphal confessed to feeling like a second-class citizen because he and Murray had to stay out of the way of the optical astronomers once the sky became dark enough to observe faint objects at optical wavelengths. For them, observing was certainly not *romantic*, which is how Maarten Schmidt described his time at the prime focus of the 200-inch telescope. Instead, the geologists were too busy working within severe time constraints and under physically demanding circumstances. Due to an abundance of simultaneous projects, the two young collaborators would zip up to Palomar Observatory, then back to Pasadena—or Murray would continue to the Jet Propulsion Laboratory to work on Mariner 4 projects. Murray later proudly added, "and we were geologists!"[209]

In the early 1960s, the two renegade geologists set out with Robert Wildey, a recent Caltech PhD, to take the temperature of Jupiter's cold atmosphere by measuring its infrared light. This required installing their bulky instruments and unruly cables inside the east-arm tube of the 200-inch telescope's support structure.

> "For quite a while I was on the outside looking in, but I had the advantage that astronomers were tolerant of our exotic experiments, whereas they are ferociously difficult with their own colleagues. The astronomers were more than prima donnas—they were egomaniacs. So, Jim and I worked quietly, like church mice over in the corner. It paid to be deliberately humble and bow a bit, because it got me on the biggest telescope in the world. Don't get me wrong. Some of my best friends are astronomers—I would let my daughter marry one."
>
> (B. C. Murray, personal interviews with the author, 2008–2010)

The slippery, cylindrical metal tube attached to the massive horseshoe—feeling a bit like a cramped submarine—is inclined 33.4 degrees to the horizontal. Inside, a staircase on wheels could be locked into position or rolled around. In the middle of the 5-foot-wide tube was an optical bench. Laden with instruments, the bench was designed to maintain its horizontal surface by tipping and tilting while the telescope slewed and tracked. Some observers spent months of their lives inside the arm, seated on one of two wooden chairs that featured custom-made back legs sawn off to fit the profile of the stairs. A wobbling mirror set in place to reflect light into the east arm would chop back and forth and make 128 steps across Jupiter, then 128 more, continuing across the rotating planet's disk. The telescope would then scan in the opposite direction to build up the image. If all went well, a completed line image of Jupiter would appear after 3 or 4 minutes.

There was no software to precisely direct the Big Eye to follow the motions of Jupiter, since the telescope's automatic tracking was engineered to synchronize with sidereal motions, not planetary motions. Hence, the geologists had to manually drive the telescope back and forth across the planet's disk. This produced unintended consequences in the mapping of Jupiter's brightness temperature: instead of being aligned with banded structures visible on the planet's surface, the contours were aligned with the direction of the scans themselves.[210] Even so, there was a prominent piling up of brightness contours near the edge of the planet's disk, indicating a darkening at the edge. This signaled that the temperature decreased with atmospheric height, as expected. Yet new infrared observations made in 1972 led to the discovery of a temperature inversion in the upper atmosphere.[211] A greenhouse mechanism was at work in Jupiter's atmosphere!

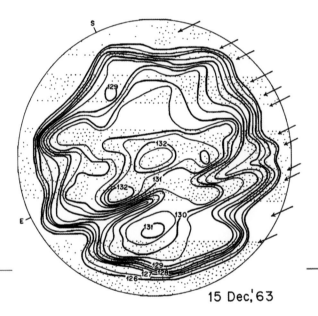

A rudimentary contour map of the near-infrared (8–14 micron) emission across the Jovian disk, observed on December 15, 1963. The scans, whose locations are shown by the arrows, produced contours at odds with the band structure, marked by dots. The band structure was visible on photographs taken simultaneously with a 35 mm camera attached to the detector. Planetary east and south are marked on the edge of the disk. Figure 4 from R. L. Wildey, B. C. Murray, and J. A. Westphal, "Thermal Infrared Emission of the Jovian Disk," *Journal of Geophysical Research* 70 (1965): 3711. Copyright by the American Geophysical Union.

Chapter 11

Jupiter's Hotspots

Jim Westphal was not afraid to take on any kind of challenge, no matter how far afield it was, sometimes radically changing focus as he did so. When another research group announced in 1969[212] that excess infrared radiation (that is, of a thermal nature) was flowing out of Jupiter, he was eager to investigate the phenomenon with the 200-inch telescope. When he scanned across the face of Jupiter with a small-aperture photometer to see how the light was distributed at a wavelength of 5 microns, he saw a jagged band of enhanced radiation. Visual-light photographs taken immediately afterward showed a wealth of detail, including some very small, intensely dark spots embedded in lighter regions. These spots—primarily in the equatorial regions—corresponded to the locations of the spots that were bright at 5 microns. Furthermore, the temperature of the spots was hotter than could be explained by reflected sunlight alone. Westphal suggested that the dark regions were holes in the clouds through which he could see into Jupiter's warm interior. He published his data in 1969,[213] although at the time some of his colleagues did not believe his findings.

It took an innovation referred to as "video pictures" to help sort out the complex features seen in Jupiter's atmosphere. Westphal, Keith Matthews (a Palomar instrument scientist), and Richard Terrile (a Caltech graduate student) recorded hundreds of high-resolution images of Jupiter at 5 microns. They took a video image every 3 minutes, immediately followed by a photograph taken with familiar Kodachrome film. Juxtaposed, the 5-micron images and color photographs helped them decipher areas of different chemical and physical states, as well as distinct cloud levels in the atmosphere.[214] The brightest regions at 5 microns represented the deepest observable layers in the planet's atmosphere and appeared blue-gray,[215] while the white zones and Great Red Spot were the coldest regions. Their time-lapse videography revealed large, rapid variations in 5-micron flux over extensive areas of the disk. Were clouds moving, condensing, and evaporating in Jupiter's atmosphere?

An unexpected opportunity for a close-up view of Jupiter's atmosphere arose in the mid 1970s, as the Jet Propulsion Laboratory prepared for two flyby Voyager missions. Due to its state-of-the-art infrared instruments and detectors, the 200-inch telescope was engaged to help locate targets of scientific interest. In return for astronomers' efforts, the Voyager missions promised to capture high-resolution multicolor data, including spectroscopy, of individual 5-micron hotspots, the smaller of which are about the size of North America. But there was a catch. The complex sequence of spacecraft commands required that target selections be locked in 30 days before the spacecraft reached Jupiter.

Unfortunately, the locations and activity levels of hotspots tend to be fickle, with some features enduring for several months, while others form and dissipate over mere days. Even on Earth, meteorologists cannot make accurate four-week forecasts of storms.

> **NOT SO NICE:** In the pioneering days of infrared astronomy at the 200-inch telescope, the bulky and temperamental instruments were operated manually either from inside the east arm or from the Cassegrain cage bolted to the bottom of the tube structure. Observers would remain with their instruments for the duration of the observations, shivering in darkness amidst ice-cold metal protrusions. In the Cassegrain cage, the faculty advisor would occupy the only seat available in the 12 × 12 × 6-foot chain-link enclosure. His graduate student would hang by straps—like a spider—and be wedged in with pieces of foam to keep from falling or crashing to death on the sharp metal corners of instrument casings in the middle of the night.[216] In spite of this, graduate students would each spend hundreds of hours over four years acquiring data for their dissertations—and retain extraordinarily fond memories of Palomar. Sitting under the whirring telescope and rumbling dome during long hours of observing was often described as a "most relaxing job." During that time, faculty advisors would impart their philosophies of life and science, thus shaping their students' research and personal lives forever.

In September 1978, five months prior to Voyager 1's first encounter with Jupiter, Westphal and Terrile began frequent monitoring of the planet with the 200-inch telescope.[217] They had updated their single detector to a new 128-pixel linear array—a by-product of the Vietnam war. Their imaging process, sped up more than a hundredfold, recorded significant brightness changes in the warm belt regions. Interpreted as the clouding over and clearing of regions in the upper atmosphere, these were the first hints of an active Jovian world harboring not-so-friendly clouds.

Chapter 11

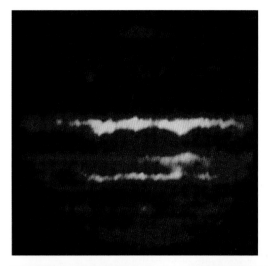

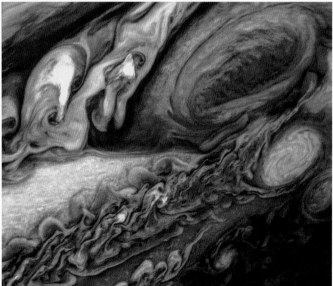

Ground-based infrared images and concomitant images from space provided direction to the approaching Voyager 1 flyby, and insight into Jupiter's rambunctious atmosphere. Top left: A 5-micron image taken at the 200-inch telescope on January 10, 1979, is shown in false color: the hottest areas are white, the warm areas red, and the coldest areas black. Top right: A visible-light image taken on the same day from Voyager 1 reveals changing brightness, interpreted as large-scale clouding-over of some regions and clearing of others. The Great Red Spot is visible in both images. Bottom: Jupiter is a fluid planet with an enormous heat capacity. The iconic Great Red Spot—a warm ring surrounding a cold central area—was imaged by Voyager 1 from 5,000,000 km away on March 1, 1979. The Red Spot is twice the size of Earth, and the winds at its outer edge roar at 100 meters per second. The smallest features that could be resolved were around 95 km across. Top left and right from figure 3 from R. J. Terrile et al., "Infrared Images of Jupiter at 5-Micrometer Wavelength During the Voyager 1 Encounter," *Science* 204 (1979): 1007; bottom courtesy of NASA/JPL.

As Voyager 1 approached Jupiter, the subdued, pointillistic images taken from Earth were replaced by images of strongly patterned, psychedelic structures. They revealed a dynamic, almost frightening world of storms, violence, and clouds with huge height variations and dramatically changing forms. Although Westphal, Terrile, and their colleagues had first glimpsed Jupiter's intricacies from Palomar, the reality from space struck them like a "punch in the face."[218]

Andy Ingersoll, a planetary scientist on the Voyager imaging team, was caught up in the thrill as the data came flowing in. He had been paying close attention to the findings of Westphal and Terrile—especially their claim that Jupiter was much more active than people had assumed. Ingersoll hoped that the arrival of Voyager 1 at Jupiter would help him solve a nagging dynamical puzzle posed by the Great Red Spot. His intuition had told him that the region surrounding the Spot should be quiescent. Otherwise, how could the Spot have lasted more than 300 years? Surprisingly, the region turned out to be chaotic—churning and bubbling and roiling—which only deepened the mystery of its long-lived storms. Westphal and Terrile already knew about the chaos from earthly observations, but Ingersoll had to see it up close with Voyager 1. Something extraordinary was going on, something beyond human experience.

From Frying Pan to Fire

It was beyond Jim Westphal's wildest dreams that, 26 years after he had discovered hotspots on Jupiter, a man-made probe launched from the Galileo spacecraft would descend by parachute into one of these spots. On December 7, 1995, after a fiery entry into Jupiter's upper atmosphere, the probe deployed its 8-foot parachute and began measuring the planet's gaseous composition with an onboard mass spectrometer. Although most of Jupiter is covered in clouds, the probe happened to dive into a hole, where some of Jupiter's central heat gushes out. For 57 minutes—before the clouds became opaque—the descending probe transmitted data that revealed turbulent atmospheric conditions. After traversing 156 km (97 miles) in altitude, the probe fell silent as the outside pressure reached 23 atmospheres and the temperature reached 426 K (307 F). It must have been crushed long before it could enter the metallic hydrogen layer at Jupiter's core, sweltering at nearly 30,000 K.

Remarkably, many of the earlier inferences made from Palomar data were confirmed. Specifically, as the Galileo probe descended into the hotspot, it encountered extremely high winds, unexpectedly dry conditions, strong turbulence, and major

Chapter 11

temperature variations, as had been inferred in 1969 by Westphal during his pioneering observation. The probe's direct measurements of the hotspot at 273 K (32 F) were close to Westphal's relatively crude estimate of 300 K (80 F, slightly above earthly room temperature). As the temperature of the surrounding regions is cooler, around 143 K (−200 F), Westphal was right: hotspots are simply holes in the clouds of Jupiter's atmosphere through which heat from the interior escapes.

A hotspot, colored royal blue, sits in the middle of this hypothetical view of a region between Jupiter's cloud layers. The artist's rendering—based on CCD images taken by the Galileo spacecraft—looks toward the horizon and spans a 34,000 km × 11,000 km region near the Galileo probe's entry. Like the matching piece of a puzzle, the white plume in the foreground, which rises up tens of kilometers, corresponds to one of the holes visible in the upper cloud layer. The blue streaks in the lower cloud layer represent a circulation pattern of dry air that sinks into the hotspots and clears away the clouds. The thin haze that remains above the hotspot had been inferred by Richard Terrile from his Palomar data. (R. J. Terrile, "High Spatial Resolution Infrared Imaging of Jupiter: Implications for the Vertical Cloud Structure from Five-Micron Measurements," PhD thesis, California Institue of Technology, 1978, 1.) Credit: NASA/JPL.

However, the Galileo probe's drop into Jupiter's atmosphere at the end of 1995 was a mere tickle compared to the weeklong assault on a cosmic scale that had occurred two years earlier.

Lord of the Planets

Colossal Jupiter, around 300 times the mass of Earth, is the solar system's bully toward comets that venture in from deep space. Its strong gravitational attraction can sling some comets around, converting their parabolic orbits into short-term orbits, while it can grab hold of other objects on short-term orbits and eject them from the solar system. In practice, Jupiter protects the inner planets from frequent impacts with asteroids—bodies affectionately called "solar system vermin" when they occasionally blaze trails across astronomers' photographic plates.

Around 1929 Jupiter captured a comet so firmly that the planet itself, instead of the sun, became the focus of the comet's orbit. Then in 1992 the small comet looped around too close to Jupiter—well inside the Roche limit, where tidal forces become insurmountably large. The fragile comet body strained against these forces until it broke apart into fragments that spread out like a string of pearls along its orbit, as later reconstructed by astronomers. Eight months after this close passage, the fractured comet was discovered by a trio of well-known comet hunters who were photographing the sky with the oldest and smallest telescope at Palomar.

Fascinated by collisions and impact craters—both terrestrial and extraterrestrial—astrogeologist Eugene Shoemaker divided his time between the United States Geological Survey, teaching planetary science at Caltech, and searching for stray bodies that might smack Earth.[219] Until 1973, there were no systematic efforts to track potentially menacing solar system bodies, nor were any statistics available on how often impacts occur. True pioneers, Shoemaker and his sharp-eyed wife Carolyn began searching thousands of paired photographs taken a day or so apart with the 18-inch Schmidt. They were hoping to find slight shifts in the position of an object against the relatively fixed distant stars, the telltale sign an object is moving within the solar system. In 1989, the Shoemakers welcomed amateur astronomer David Levy into their team, and the trio worked in harmony through the long laborious hours of observing.[220] On one of the pairs of small-scale photographs taken with the 18-inch Schmidt camera, a shifting object appeared strangely elongated, like a comet. When the comet's orbit was computed, astronomers across the world were stunned to learn that the broken comet was on a fast-track collision course with Jupiter. The fragments of the doomed comet, officially named Shoemaker-Levy 9, would vividly demonstrate one of the solar system's basic physical processes: accretion by impact.

Chapter 11

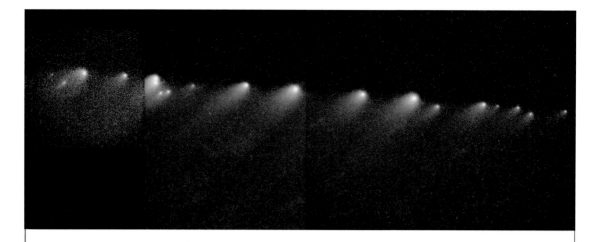

What was originally the single body of comet Shoemaker-Levy 9 appeared as a string of fragments—some a few kilometers in diameter, some smaller—racing toward an energetic, high-velocity collision with Jupiter. Extending over a distance of a million kilometers, the fragments were assigned letters of the alphabet corresponding to their predicted order of impact. Some fragments later merged, split, or even disappeared. Credit: NASA, ESA, H. A. Weaver and T. E. Smith (STScI).

Incoming!

Gerry Neugebauer, in the final year of his Palomar Observatory directorship, preemptively canceled the regular observing schedule during two weeks in July 1994, the predicted window for the collision. The astronomers assigned to record the event faced a curious situation. No one knew the exact times of the impacts, what to expect, which instruments would be best, or even where the bulk of the radiation would come out—in the visible, near-infrared, or thermal infrared. Furthermore, the cometary fragments were expected to hit the far side of the planet, and earthly observers would have to wait until Jupiter's rotation carried the wreckages into view.

Instrumental resources were pooled by a team of astronomers, led by Phil Nicholson of Cornell University, that included Neugebauer and Keith Matthews of Caltech. Matthews, being a virtuoso instrumentalist, had figured out a seemingly magical way of using the 200-inch telescope's chopping secondary mirror—widely considered a piece of technical artistry—to rapidly flash the image of Jupiter alternately to two chosen instruments. All this would be synchronized with intermittent measurements of the sky background. By now, the 200-inch telescope had a large pool of infrared instruments, including a new near-infrared camera with a suite of broadband and circularly variable filters built by Matthews. Since Jupiter's spectrum shows strong methane absorption bands around 2.3 microns, the new camera—sensitive to this wavelength region—was expected to darken the planet's surface and enhance the visibility of columns of hot gas created by the impacts.

For broader coverage, Cornell University contributed Spectro-Cam 10, a spectrometer capable of imaging mid-infrared wavelengths and of taking spectra from 8 to 13 microns. The dual instrument could image and take spectra of gases that had been heated to room temperature by the impact. Both instruments were simultaneously bolted to the back of the sturdy battleship of a telescope. Since observations were made in the infrared, the dome was opened as early as 1 o'clock in the afternoon, with the sun still up. Observatory personnel were stationed around the dome's catwalk to warn the telescope operator should the sun threaten to shine into the dome and directly onto the mirror.

During this hectic two-week period, Palomar was part of an extensive network of ground-based telescopes located in South Africa, Europe, North and South America, Hawaii, Japan, Australia, and Antarctica, all aimed at Jupiter. As the string of fragments impacted Jupiter, electronic messages flew back and forth between countries, broadcasting new sightings and updating ephemerides. On July 21, 1994, the astronomers at the 200-inch telescope learned that, to detect one of the cometary impacts, the telescope would have to be pointed at Jupiter as it first rose above the horizon. However, to protect the telescope from being accidentally driven into the ground, or the 14.5-ton mirror potentially dropping out of its cell, several levels of clever software were in place that required overrides for such situations. The astronomers wanted the telescope, already parked at 10 degrees above the horizon, to be inclined even lower. They implored the director to permit them to override the software stops. He relented and the telescope was lowered to a mere two degrees above the horizon, where its motion was stopped by a hotwired switch—*clap!*—that said *no further*.

> Eager to record the fast-approaching impact of a Shoemaker-Levy 9 fragment, the observers at the 200-inch telescope rallied the director to break the rules and violate the safety stops. They shouted, "Yeah, to hell with the stops!" It worked. The fragment was detected just as Jupiter rose above the Earth's horizon.
>
> (P. Nicholson, personal interview with the author, 2009)

Chapter 11

Bolide Flashes and Splashes

The entry of the fragment designated R into Jupiter's atmosphere was first perceived as a blip in the infrared. The astronomers reckoned that the source was a few hundred kilometers above the denser atmosphere, but still behind Jupiter's limb. A minute later, a pillar of hot atmospheric gas heated by the impact spurted 3,000 km above the limb of the planet, producing a second, much brighter glow that lasted 20 to 30 seconds. Although the impact site itself was still on the backside of Jupiter, the thermal pillars extended high above the planet's limb, into Earth's view and visible in the sunlight. The pillars radiated strongly in the thermal infrared. In the 200-inch data room, the mood was one of hushed excitement and whispered exclamations. There was a lull. Then, to everyone's astonishment, about 10 minutes later a third flash generated a 600-fold brightening of infrared flux that saturated the monitor. The glowing column of hot gas—composed of both cometary and dredged-up Jovian material—had crested and was crashing back into the atmosphere. In doing so, it shock-heated atmospheric gas to several hundred degrees Kelvin. The impact site extended more than 8,000 km along Jupiter's limb, now carried into full view by the planet's rotation.

During this time, astronomers in the data room frequently tweaked the instruments to capture the bulk of the flux from the event. This was a complex and indeterminate task as the shape, dimension, and viewing geometry of the impact site kept evolving over time. The challenge was to keep the brightest emission spot centered on the assortment of instrumental apertures by adjusting their positions and sizes. By happy coincidence, the sensitivities of the infrared detectors matched the peak of thermal emission from the pillars and from the hot remnants. The astronomers in the control room were relieved.

For some weeks afterward, observations continued in the near-infrared as the impact scars blurred.[221] The amount of blurring allowed astronomers to estimate wind speeds in Jupiter's stratosphere. Initially, the collisions had held the promise of a unique opportunity to peek below Jupiter's cloud layers as lower-atmosphere gas would be dredged up with the erupting material. However, only gas of normal composition was detected in Jupiter's stratosphere.[222] Since the comet fragments turned out to be somewhat less massive than had been predicted by Gene Shoemaker, they did not penetrate as deeply as expected. Most of the gases seen in the plumes, especially water vapor, are now thought to have come from the comet itself, not from deep Jovian clouds. "Seismic" oscillations in Jupiter's atmosphere—indicating deeper impacts—were not detected.

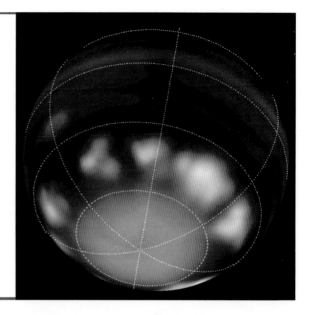

The impact of comet Shoemaker-Levy 9 with Jupiter generated multiple bright fireballs, each roughly Earth-sized, that encircled the planet's southern hemisphere. Here, heated particulate debris—tossed high into the Jovian atmosphere during impact—appears to glow against the dark planet. Infrared images were taken with the 200-inch telescope on July 23 and 24, 1994, over a period of several hours while Jupiter's rotation rapidly carried nine impact sites into view. This artful composition is a computer-generated mosaic of those separate images, each projected onto a map grid, then wrapped onto a globe. The Great Red Spot is the faint oval above and to the left of the impacts. Credit: NASA/JPL-Caltech.

The annihilation of a comet by a planet not only excited—and even entertained—the world with its fireworks. It also triggered a paradigm shift away from the belief that collisions between solar system bodies are rare events. Although the initial violent buildup of the planets through collisions and accretions is long past, the process clearly continues at a reduced level. Even Earth could be knocked—and seriously damaged—by a sizeable piece of rubble such as the Shoemaker-Levy's fragment G. Gas giant Jupiter had just demonstrated that its powerful atmospheric circulation has the capacity to disperse impact structures until they span thousands of kilometers. By contrast, if even one of the smaller cometary fragments had hit Earth, it would have burst into a giant fireball, leaving behind massive destruction and loss of life.

An example of a potentially threatening object is the fast-moving, so-called "Earth-crossing" 1.6-km-wide asteroid named Icarus, discovered by Walter Baade with the 48-inch Schmidt camera on June 26, 1949.[223] Nineteen years later, Icarus sped close past Earth again, grazing it within a few million kilometers—a shockingly small distance by cosmic standards. Another time, the result might be a collision with Earth. Planetary astronomers have, of course, long known that understanding the physical and orbital properties of comets and asteroids is crucial for estimating any possible future threats to Earth. In response to Congressional interest, NASA initiated a formal program to detect and track bodies that might one day collide with Earth. But perhaps scientifically more interesting is the question of how and where these objects originate.

Chapter 11

Solar System Refrigerators

Comet Shoemaker-Levy 9 could have broken loose from either of two conjectured repositories in the outskirts of the solar system: the Kuiper belt or the more distant Oort cloud. Both are vast regions believed to be filled with orbiting rubble left over from the assembly of the solar system's planetary disk. Far from the sun, these two chilly repositories effectively serve as grand refrigerators that have preserved their contents fresh and uncontaminated for 4.6 billion years—the entire lifetime of the solar system. The Kuiper belt remained merely a hypothesis until 1992, when David Jewitt and Jane Luu—using digital cameras and sophisticated computer software—detected the first frozen body moving beyond the orbit of Neptune.[224] The door to the nearest refrigerator had been opened just a crack.

> We are reminded that "Earth is a member of a family of planets and that planets occasionally are struck by the interplanetary wanderers called comets and asteroids."
> (E. M. Shoemaker, "Comet Shoemaker-Levy 9 at Jupiter," *Geophysical Research Letters* 22 [1995]: 1555–1556)

Down the road from the University of California at Berkeley, graduate student Mike Brown was living on a sailboat in the Berkeley Marina while he studied comets and the Jovian magnetosphere. When Comet Shoemaker-Levy 9 was discovered, he saw his two thesis topics collide at 90 km/s. "Happily," he jokes, "I didn't work on asteroids and Earth!"[225] He became interested in exploring the Kuiper belt through Luu, who was then a UC Berkeley postdoc. Hired in 1996 as a planetary astronomer at Caltech, Brown learned that the Palomar 48-inch Schmidt camera was just finishing the second-generation Palomar-National Geographic Sky Survey and would become available for new tasks. In a "lightbulb moment," he realized that the Schmidt camera—with its large field of view, a generous allotment of observing time, and the availability of two mountaintop observers every dark night—would be the ideal instrument with which to search for more icy bodies in the Kuiper belt.

Whereas the deep, narrow-field searches by others had yielded hundreds of small, faint pieces of debris, the rare-but-substantial Kuiper belt objects were thought to be the top players in the dynamics of the outer solar system. Brown hoped to find these coveted bright objects, study their surface composition, and constrain theories of their formation. However, after two years of intense observing with the Schmidt and an additional year of follow-up, the grand total of new Kuiper belt candidates in his arsenal was, alas, *zero*.

The indefatigable Brown was convinced that this was the right telescope, but it needed a more sensitive detector than photographic plates. Fortunately, by the late 1990s, Eleanor "Glo" Helin, a highly respected maven of Earth-crossing asteroids and comets, had already been angling for just such a detector. Having struggled for 25 years with photographic films and plates at the 18- and 48-inch Schmidt telescopes, Helin managed to convince her employers at the Jet Propulsion Laboratory to upgrade the 48-inch Schmidt camera. The system of recording data on inefficient, cumbersome photographic plates was replaced with a 48-million-pixel mosaic of charge-coupled devices (CCDs). Instead of the meager 1% efficiency of the photographic emulsion, the new CCDs recorded up to 90% of the incident photons and detected objects 10 times fainter. Helin began a new survey called Near-Earth Asteroid Tracking[226] and cataloged more than 26,000 new objects. Brown, former doctoral student Chad Trujillo from Gemini Observatory, and David Rabinowitz from Yale University harnessed the new CCD camera for another round of their search for objects more distant, and potentially larger, than Pluto. It took them just one month, and they were able to record objects ten times fainter than with photographic plates.

Pieces of rubble in the outer solar system are so far away from Earth that they appear as small, faint smears of light in a field of stars. Whereas light emitted by a star dims as the inverse square of its distance from Earth, an object that shines only by the reflected light of the sun dims with the 4th power of its distance from the observer. This is because the inverse-square law applies to the total distance the light travels: from the sun to the object, then from the object to Earth (roughly the same distance again, for objects in the far reaches of the solar system). Hence, as the object's distance doubles, the light dims by a factor of 16 (2^4). The identification of distant rubble is further hampered by another coincidence of nature. A small object covered with highly reflective ice reflects the same amount of light as a large one covered with light-absorbing dirt, and the two kinds of objects are nearly indistinguishable at large distances. Rubble far from the sun is also very cold, around 40 K, which means that its thermal radiation peaks at about 75 microns, far beyond visible or near-infrared light.

In June 2002, Brown and Trujillo discovered an object more distant than Pluto and about half its size.[227] Its orbit, tipped eight degrees off the ecliptic, yet nearly circular, was considered bizarre at the time—even a fluke. Yet Quaoar, as the object was eventually named, was characteristic of a burgeoning new class of objects: icy planetoids in the outer solar system. Quaoar also turned out to be one of the first nails in the coffin of Pluto's designation as a planet.

CHAPTER 11

Fragments from the Oort Cloud

Searching farther and farther away from the ecliptic, in the fall of 2003 Brown, Trujillo, and Rabinowitz discovered the first object well beyond the edge of the Kuiper belt, naming it Sedna.[228] Immediately recognized as a strange world, Sedna is half the size of Pluto and has a deep-red surface which appears neither dark and rocky nor bright and icy. Scientifically, it was an unexpected but important find: It roams a dynamical region of the solar system formerly believed to be empty. In fact, Sedna is dynamically distinct from the entire population of known Kuiper belt objects. It takes about 11,400 years to complete its enormous, highly eccentric orbit, which is tipped eleven degrees off the ecliptic. Even during its closest approach to the sun (at perihelion, in astronomers' parlance), at 76 AU, Sedna is too far from the known planets to be significantly influenced by their gravitational pull. When Sedna is at its farthest from the sun (at aphelion), at about 1,000 AU, it is not yet far enough away to be subject to orbital disruption from stars passing by the solar system. What form of dynamical violence, then, did it suffer to settle into such an extreme orbit?

Perhaps a massive rogue planet kicked Sedna from the Kuiper belt (at 30 AU to 50 AU from the sun) to the inner Oort cloud as it plowed through the solar system and out again. Or Sedna's perturbed orbit could be explained if the solar system formed inside a cluster of stars. If a nearby star had skirted the infant solar system, allowing the sun to capture Sedna and other icy bodies from its grasp, might Sedna even be the nearest object of extrasolar origin?

In any case, Brown was convinced that finding more fossils similar to Sedna would lead to an understanding of how they formed, as well as how the solar system assembled. He searched diligently for the next five years, switching from the Palomar 48-inch Schmidt camera to the 200-inch telescope to try to catch the very faintest ones. In 36 nights of point-and-shoot observing, he found no further candidates. Yet Sedna was there, demanding that planetary astronomers take another look at the vastness and richness of the embryonic solar disk's contents.

Solar System Shuffle

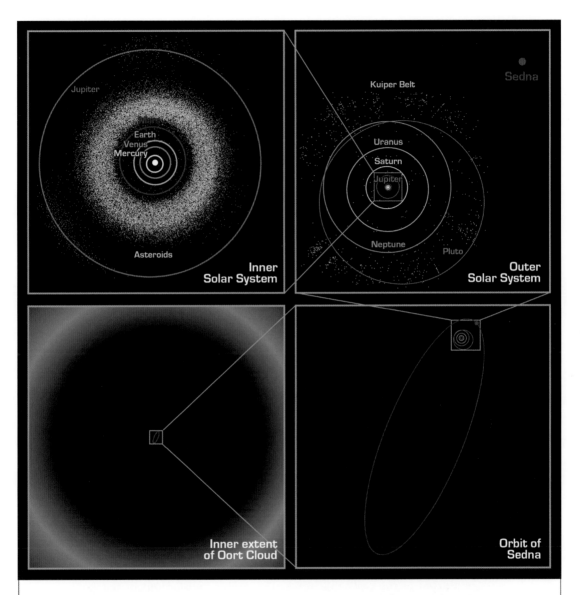

This four-panel schematic, to be read clockwise from the upper left, pictorializes the immense scale of the solar system as it is currently understood. The spherical Oort cloud traces the gravitational boundary of the solar system. As with the asteroid belt and Kuiper belt, it is believed to be filled with icy rubble left over from the assembly of Earth and the major planets 4.6 billion years ago. The Oort cloud is thought to extend from about 20,000 AU from the sun (500 times farther away than Pluto) to about 200,000 AU (about halfway to the nearest star). Image courtesy NASA/Caltech.

Chapter 11

THE GHOSTLY OORT CLOUD: The existence of a distant cloud of icy rogue comets was hypothesized in 1950 by Dutch astronomer Jan Oort (J. H. Oort, "The Structure of the Cloud of Comets Surrounding the Solar System and a Hypothesis Concerning Its Origin," *Bulletin of the Astronomical Institutes of the Netherlands* 11 [1950]: 91–110) in order to resolve the following paradox. Fragile comets had been observed streaking through the solar system on unstable orbits diving in from far away. Besides having their volatiles boil off near the sun, their numbers tend to dwindle as they suffer captures, collisions, and sometimes ejections. Yet comets are known to have been traversing the solar system since its formation, so there must be a source for frequent replenishment. But where? Oort conjectured that icy debris swarms in a roughly spherical shell far from the sun, in a kind of cold "safe house." The only concrete evidence for the existence of this Oort cloud is that each year, some pieces of debris plunge toward the sun on nearly radial orbits. As they approach the sun, they develop long gas and dust tails, appearing to us as comets. Models now predict that there may be as many as a trillion icy objects populating the Oort cloud.

A year later, in 1951, astronomer Gerard Kuiper (G. P. Kuiper, "On the Origin of the Solar System," *Proceedings of the National Academy of Sciences* 37 [1951]: 1–14) suggested that some comet-like debris from the formation of the solar system should orbit much closer to the sun, in a belt just beyond Neptune. Otherwise, he reasoned, the original planetary disk would have had a physically implausible sharp boundary. His hypothesis was reinforced in the early 1980s, when computer simulations of the solar system's formation predicted that a disk of debris should surround the system beyond Neptune, the last of the gas giants.

Tallying Pieces of the Puzzle

Continuing their searches, Brown and his two collaborators discovered in December 2004 an especially enigmatic and bizarre object that they named Haumea.[229] Intrinsically and visually brighter than any object they had found so far, it had odd characteristics that defied a simple explanation. Spinning end over end once every four hours, Haumea was the fastest-rotating gravitationally bound object known in the solar system. It resembled an iceball shaped like a deflated football and it had two moons!

It took Brown and his team three years to figure things out: Haumea was originally a spherical body nearly the size of Pluto, with a high-density rocky core and an overlying low-density ice mantle. Then, early in the life of the solar system, Haumea must have been struck a glancing blow by a fast-moving object that cracked open the icy mantle and stripped it away. Left spinning from the collision, Haumea's rocky body was sculpted and elongated over time. Two small fragments of the mantle went into orbit around Haumea as moons, and the orbital dynamics of the larger moon allowed Brown and his team to determine the mass of Haumea and of the moon. The density of Haumea was determined from its degree of flattening compared to a sphere (flatter meaning less dense, and fatter more dense). Thus, Haumea appears to be made of mostly rock surrounded by a thin layer of the ice remaining after the collision.[230]

The smoking gun for such a collision came from new evidence acquired by serendipity—"the kind you prepare really hard for," according to Brown.[231] Hoping to make sense of the spectroscopic and dynamical data they had collected for more than two dozen Kuiper belt objects, Brown and his collaborators plotted the distance from the sun for each orbit, how much the orbit is inclined to the disk of the solar system, and the orbital eccentricity. The degree of iciness of each object was represented by a color: black (not icy), gray (a bit icy), and white (pure ice). Out of 30 objects scattered randomly about the Kuiper belt, only six were uniquely pure ice.

It was midnight in Sicily when Brown and Kristina Barkume, his graduate student, pored over the plots as she was preparing a presentation for a conference the next day. Her talk was supposed to be about why the six objects were so icy, but she hadn't come to a conclusion yet. She was pointing out that their orbits seemed to match Haumea's orbit and wondering why that might be so when Brown suddenly blurted out: "Oh my god, these are pieces of Haumea itself!" The two realized that, even though the six objects aren't physically close to each other, their *orbits* are essentially identical. Since the six icy bodies resembled the moons of Haumea, they were likely part of the family of icy debris knocked off in the same collision. However, instead of going into orbit around Haumea, the six went into orbit around the sun. When calculations showed that there is a one in a million chance that the similarity of their orbits was random, it was a pure *aha!* moment.

In their 2007 *Nature* paper entitled "Discovery of a Collisional Family in the Kuiper belt"[232] Brown, Barkume, and two more graduate students presented the collision as evidence that catastrophic collisions can spew debris across the solar system. The cores of the giant planets may have formed from such collisions and, over time, some fragments and shards may have plunged toward the sun. Perhaps in the past a few such shards may even have collided with Earth.

Anatomy of a Murder: The Demise of Pluto as a Planet

Sometimes in science, a few examples are good enough to formulate a hypothesis or conclusion. But every Kuiper belt object was different enough that Brown and his team wanted to search for yet more distant objects before formulating theories or drawing major conclusions. In January 2005, they reanalyzed the data acquired in 2003, this time setting the automatic search criteria to flag objects with smaller transverse motions, since objects appear to move more slowly across the sky the more distant they are. A new object that had been below the sample cutoff in 2003 was flagged, and they named it Eris.[233] Images and spectra taken with the Hubble Space Telescope and the Keck telescope revealed an icy, rocky body covered with a layer of methane that presumably seeped from the interior and froze on the surface. As on Pluto, the methane had reddened through chemical reactions due to sunlight. As Eris was 2,400 km wide—distinctly bigger than Pluto—and 30% more massive, it appeared to be its larger twin. Was it perhaps the long-sought tenth planet?

Soon thereafter, in March 2005, Brown and his students discovered Makemake, an icy object three-quarters the size of Pluto and also covered in methane ice.[234] Was it therefore also a planet? For a long time, Pluto had seemed to be unique, a relatively small misfit orbiting the sun on an inclined, strongly eccentric orbit that sometimes carries it inside Neptune's orbit. Not only that, Pluto's rock-and-ice composition starkly contrasts with the compositions of Uranus and Neptune, its neighboring gas giants. When there is only one object, there are no checks and balances to guide the development of theories.

Now there were three similar objects to compare: the distant Eris, the relatively close Pluto, and Makemake in the middle.

Were the largest Kuiper belt objects to be classified as planets? Was Pluto the link between a continuum of objects that begins with Mercury and ends with the likes of Sedna? Either the solar system contained a vast population of planets, or Pluto had been misclassified as a planet in 1930 and had now to join its kindred spirits in the Kuiper belt. After intense debate, some on emotional and some on intellectual grounds, the International Astronomical Union decided in 2006 that Pluto must leave its planetary burrow and join a new type of solar system population named "dwarf planets" or—really—large Kuiper belt objects.

> "Though Eris is the largest object—and its discovery led to the demotion of poor Pluto after all this time—it is not the most interesting one. Sedna and Haumea are the superstars. They are scientifically rich, whereas Eris was just a big hammer to beat Pluto to death. Which is good: somebody needed to do it!"
> (M. Brown, personal interviews with the author, 2008–2019)

The reconceptualized solar system now consists of the four rocky terrestrial planets and thousands of asteroids out to around 4 AU from the sun; the four gas giants out to Neptune at 30 AU; and a vast disk-shaped population of icy rubble and boulders—including Pluto—out to 50 AU. The latter is the Kuiper belt. Beyond this solar system lies the conjectured comet-bearing Oort cloud—with Sedna occupying its innermost region, and the outer region extending to more than 100,000 AU from the sun.

Seeking Planet X: Hindsight

After Pluto was discovered in 1930 and named the ninth planet, astronomers began to search for the elusive and perhaps more massive Planet X. From 1961 until 1985, among Charles Kowal's main assignments as a Mount Wilson and Palomar observer was to search for supernovae in other galaxies. However, under the radar Kowal made time to systematically survey the solar system to the depths that could be reached with the 48-inch Schmidt camera. He searched 6,400 square degrees of the sky—the entire ecliptic region—for distant, slow-moving solar system objects and, mainly, for the mythical Planet X. It was tedious visual labor to inspect the mountains of photographic plates he acquired, but Kowal had the ideal temperament. In October 1977 he spotted an object in the region between Saturn and Uranus, which everyone had thought was empty.[235] Hopeful that he had discovered Planet X, Kowal named the object Chiron, which embodies a kind of mythological uniqueness.

Alas, its orbit turned out to be unstable and chaotic, and Chiron, too, was demoted—to a population of comet-like objects. Kowal named that population Centaurs to emphasize that they are neither asteroids nor comets nor anything else known. Although Jupiter wields the dominant influence over their motions, Centaurs still lurch as they repeatedly crisscross the orbits of Saturn and Uranus. One by one, over times that are brief compared to the age of the solar system, Chiron and its brethren will be permanently ejected or demolished by a head-on collision with one of the massive planets. But finding Chiron meant that there had to be a vast, undiscovered population of replacement objects further out that supply more Centaurs. In hindsight, that should have been an early clue that the Kuiper belt existed.

Upon leaving Caltech for the Space Telescope Science Institute in 1985, Kowal systematically labeled his photographic plates and stored them in the subbasement of Robinson Hall. Years later, after Quaoar had been discovered, Brown extrapolated its orbit backward and found a match with the coordinates of one of Kowal's plates taken on May 10, 1983. Brown eagerly pulled Kowal's plate out of its envelope to see if he could locate Quaoar on it. There it was, a tiny dot among thousands of stars on the plate, serving as a strangely satisfying connection to the search for Planet X begun by Kowal more than a quarter century earlier.

Seeking Planet Nine

In 2012—a decade after Sedna was found—Chad Trujillo, codiscoverer of Eris and now at the Gemini Observatory, and Scott Sheppard, staff scientist at Carnegie's Department of Terrestrial Magnetism, found another dwarf planet. About half the size of Sedna, it orbits the sun in the inner Oort cloud. Officially tagged 2012 VP_{113}, its presence signaled that Sedna wasn't just a fluke. Brown was happy. As more such remote objects were found, Brown and colleague Konstantin Batygin, a young theoretical planetary astrophysicist, teamed up to analyze data from their respective observational and theoretical backgrounds. After a year and a half of detailed mathematical modeling and computer simulations, they found that the orbits of seven recently discovered Kuiper belt objects were bewilderingly clustered—even aligned—in physical space.[236]

Calculations showed that it would take the gravitational influence of a relatively massive object far out in the solar system to produce the observed synchronization of orbits.

In 2016, Brown and Batygin published observational constraints for the predicted orbit and for the location within the orbit of the hypothetical object, which they nicknamed "Planet Nine."[237] According to their models, a sustained gravitational torque (a twisting force) provided by the putative planet would gravitationally nudge wayward solar system objects to restore and maintain their clustered orbits. Planet Nine would probably weigh between 5 and 10 Earth masses (5,000 times Pluto's mass) and would be located about 20 times further away from the sun than Neptune. Scores of telescopes worldwide have by now been tasked with searching for Planet Nine, and astronomers and citizen scientists alike are analyzing archived data for evidence of small telltale motion of this distant world.

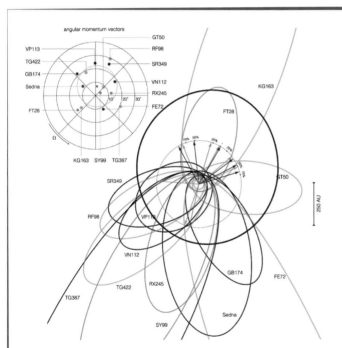

This diagram, published by Batygin and colleagues in 2019, shows the spatial clustering of the orbits of 14 recently discovered Kuiper belt objects—plus that of the hypothetical Planet Nine. The various orbits' semimajor axes range from hundreds to thousands of astronomical units, their periods are longer than 4,000 years, and they are highly elongated. The predicted orbit of Planet Nine is superposed (dark red). The orbits—viewed from the north ecliptic pole—are color-coded according to their dynamical stability. The orbits of the "Neptune-detached" objects (purple) are stable for longer than the age of the solar system, while those of objects that interact with Neptune (green) are dynamically chaotic. The orbits of other objects (gray) experience only mild fluctuations over the age of the solar system. The diagram to the upper left shows a polar projection of the directions of the angular momenta of these Kuiper belt objects. After Figure 6 from K. Batygin, F. C. Adams, M. E. Brown, and J. C. Becker, "The Planet Nine Hypothesis," *Physics Reports*, arXiv:1902.10103, February 2019: 1–92; courtesy of Konstantin Batygin.

Chapter 11

In retrospect, Palomar undoubtedly played an important role in introducing infrared observations to the study of planets. That fresh view led to the realization that a vast population of icy bodies—some with their own moons—exists in the outer solar system. Sedna informed astronomers of the existence of an extreme outer region in the solar system, and the subsequent discovery of Eris provided a physical and dynamical context for Pluto's oddities. The reclassification of Pluto as a dwarf planet changed planetary science forever: no longer can solar system objects be studied in isolation. This lesson was reinforced by the discovery of several more remote Kuiper belt objects, the synchronization of some orbits, and the putative distant Planet Nine.

Thus, the study of our solar system has spilled over into the study of extrasolar planets, called *exoplanets* for short. New discoveries there are revealing more clues about the birth of our own solar system and the spatial rearrangement of its constituents. In some sense, the Kuiper belt—thought to be left over from a time 4.5 billion years ago when icy fragments coalesced to form the outer planets—is a link between our solar system and the debris disks around other stars. How would our solar system appear from afar? How would we recognize a Kuiper belt or an Oort cloud around another star? What if we could observe an Earth-like planet at, say, 2 billion years of age instead of at 4.5 billion years (the present age of Earth)? Systems of exoplanets increasingly provide a laboratory for helping us understand the chemical, geophysical, and accretional evolution of our own solar system. Inevitably, they also introduce the question of extraterrestrial life: Are we alone? Already astronomers are asking: When we find Earth-like exoplanets, how will we go about finding evidence for biological activity on them?

Dome reflections from the newly coated 200-inch mirror. Photo by the author.

· 12 ·
Astronomical Exotica: New Frontiers

CHAPTER 12

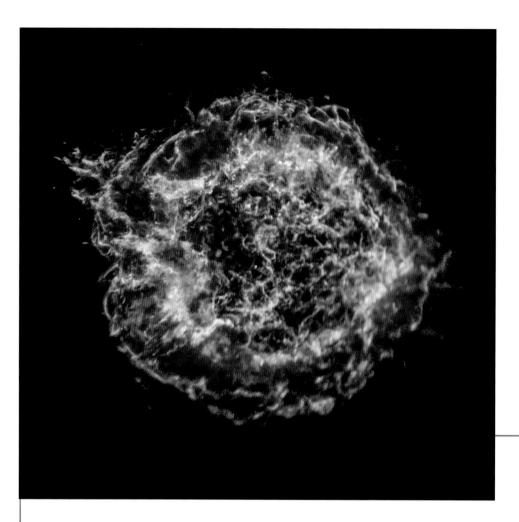

In about 1670 AD, a massive star exploded in the constellation Cassiopeia and ejected its outer layers into surrounding space. In addition to an irregular sphere of shocked gases, the image of its gaseous remnant Cas A reveals thousands of knots that travel like a hail of bullets at speeds up to 15,000 km/s—often well ahead of the shocked gas in the shell (blue). Notice also the gaseous filaments jetting out to the left and the gap in the shell to the right, both perhaps formed by two powerful jets tunneling in opposite directions early on during the explosion. At Palomar, the search for such star-shattering explosions has evolved from that of a lone Fritz Zwicky photographing the sky all night with the 18-inch Schmidt camera to the current vast international collaboration of astronomers, engineers, and instrumental specialists, who are connected by a web of electronics, computers, and software. The telescopes are now roboticized, and there is not a soul in the domes. Credit: NASA/CXC/SAO.

ASTRONOMICAL EXOTICA: NEW FRONTIERS

During the 20th century, the volume of the explored universe increased rapidly, exposing a myriad of strange beasts and processes. Sources were found to emit radiation at wavelengths ranging from the size of a proton to the size of a continent. Filamentary structures were found to stretch across billions of light-years. A strange energy was revealed that pushes everything apart. And an omnipresent form of dark matter was inferred to rule the architecture of the universe. The rapid foray into deep space was fueled by powerful new technologies developed during wartime and released to astronomers after years of sequestration by the US Department of Defense. Clever and artful application of these technologies also enabled astronomers to poke around previously inaccessible nooks and crannies of space nearby, in our own Milky Way galaxy. Among many questions, two loomed large in their minds. The first arose from an age-old human preoccupation: "Are there worlds like ours around other stars?" The second concerned the observational enigma stumbled upon in 1933 by Fritz Zwicky: "What is the nature of dark matter?" It seems counterintuitive that these two questions are linked—but in fact, the search tools developed by astronomers looking for insights into the esoteric concept of dark matter ended up casting light on the existential question of our uniqueness in the universe.

RUNTS OF STAR FORMATION

As Zwicky analyzed the velocities of individual galaxies orbiting within the Coma Cluster, it struck him that the galaxies were moving so fast that the cluster should have been flying apart. Yet the cluster appeared old, so something had to be holding its galaxies together. After considering various explanations, he proposed that there must be a significant amount of nonluminous "dark" matter within the cluster acting as gravitational glue. Although there was no way to observe this matter, Zwicky inferred that its existence was the only explanation for what he could observe.[238] This enigma was widely ignored for nearly four decades, until theorists started modeling how the disks of galaxies form and evolve. Time after time, their computer simulations of rotating disks of stars and gas showed them to be unstable and rapidly break up. In 1974, Jerry Ostriker, a flamboyant Princeton theoretical astrophysicist, proposed that galactic disks could be stabilized by embedding them in a hypothetical halo of dark matter that was three times more massive than the visible stars and gas. His idea fit hand in glove with new observations showing that stars and gas visible in our Galaxy and in other spiral galaxies were moving faster than predicted from their masses and Newton's laws.

Chapter 12

Theoreticians soon began speculating that dark matter could be made up of material ranging from diamonds to bricks to quadrillions of "tiny" (0.1% solar mass) black holes orbiting in galaxy halos. However, the prime candidate for dark matter was a large reservoir of hypothesized pitch-dark failed stars thought to be the missing link between normal stars and planets. Such runts of star formation had been postulated in the early 1960s by Shiv Kumar, a barely 20-year-old theoretical physicist working at NASA's Goddard Institute for Space Studies in New York. Kumar was curious about the structure and properties of very-low-mass stars. To figure out the minimum mass it might take to ignite a star, he modeled how interstellar gas clouds of various masses and chemical compositions collapse and fragment. As a gas cloud contracts, gravitational potential energy is released in the form of radiation and heat, which raises the temperature in its core. The contraction stops when the core temperature exceeds about 3,000,000 K, the temperature at which fusion reactions begin converting hydrogen into helium. This marks the birth of a star. It will spend most of its life on the main sequence, which is a state of exquisite balance between the inward pull of gravity and the outward pressure of hot gas heated by the energy generated in thermonuclear reactions.

But what happens if the ignition temperature for fusion is not reached? Using the tools of quantum physics, Kumar calculated that the lowest-mass clump of interstellar gas that could end up as a single main-sequence star is around 7% of the sun's mass.[239] Gas clumps of lower mass than this minimum stop contracting due to a phenomenon called "electron degeneracy." In the forming low-mass star, the electrons become so tightly packed that they cannot be jammed closer together because all the lowest energy states are occupied. Thus, as in a game of musical chairs, higher-energy electrons cannot "cool down" by squeezing into lower energy states. Instead, they continue to jiggle and vibrate, hostage to the conflict between gravitational forces and quantum forces.

As a result of electron degeneracy, these "failed" low-mass stars never get smaller than about the size of Jupiter. For the next billion years or so, they glow mostly from the release of gravitational energy accumulated during their contraction phase. Eventually, like embers of a dying fire, they cool and fade into blackness.

Initially, most of the scientific community responded with disbelief to the idea of failed stars, and, as a result, Kumar's first papers were rejected by mainstream journals. Yet for a few astronomers, the hope of discovering an entirely new class of astrophysical objects playing hard-to-get led to dogged, sometimes obsessive searches that went on for decades. These searches were guided by the assumption that failed stars should be cooler than the coolest known stars, so that they would emit most of their radiation at infrared wavelengths beyond three microns. Using the latest infrared detectors, teams began to scour stellar nurseries in the Orion Nebula and the Pleiades for their youngest, faintest, and reddest members. It seemed the sensible way to find objects whose observational characteristics were so poorly understood that no one could even predict what color they would be. Jill Tarter, a graduate student at UC Berkeley in 1975, worked on models that predicted how dwarf atmospheres and colors would appear.[240] Because dwarfs change temperature—and hence color—over time, she labeled them "brown"—a mix of many colors—to cover all the bases.[241] It then became a virtual horse race to find the first brown dwarf, with competitors from far and wide such as Rafael Robolo in Spain, Gibor Basri at UC Berkeley, and a group in Arizona that held a conference for what turned out to be a spurious signal. Then, after three decades of no-shows and false leads, the race was won unexpectedly by a dark-horse team with a unique instrument, a fresh approach, a good dose of luck, and access to Palomar's Big Eye.

THE RIGHT STUFF

The tragic explosion of the *Challenger* space shuttle in 1986 occurred as astronaut Samuel Durrance was preparing for his first scheduled flight. While NASA placed shuttle flights on hold to sort things out, Durrance took leave at Johns Hopkins University to explore the new and rapidly developing science of extrasolar planets. With that choice, he unwittingly became the first member of the dark-horse team that would assemble during the next several years. Trained as a planetary scientist, Durrance dreamed of designing and building an instrument to image faint planets, or at least preplanetary disks, as they form around nearby stars. He was joined by David Golimowski, a Johns Hopkins University postdoc whose dissertation would depend on the instrument's success. Since work on the instrument was interlaced with Durrance's astronaut duties—which included riding into space—Golimowski served as the project's anchor.

The instrumental challenge of detecting ultrafaint circumstellar material in the proximity of a bright star is the equivalent of trying to glimpse a firefly hovering inside an approaching car with glaring headlights in darkness. To block the glare of bright stars and allow surrounding faint objects to be seen, the astronaut and his student built a stellar coronagraph, an instrument with a small reflective disk in its focal plane that occults a star's light and prevents its glare from overwhelming surrounding faint objects. However, Durrance and Golimowski improved the classic coronagraph with an innovative twist. They designed a simple "adaptive optics" component that removed—in real time—the smearing of celestial images due to atmospheric turbulence and telescope tracking errors. Instead of the blocked starlight being discarded, the reflective occulting disk fed light from

Chapter 12

the bright star into an image motion sensor which detected the motion due to atmospheric fluctuations. Every 10 milliseconds, an actively controlled tip-tilt mirror corrected for the slow wandering of images.[242] This "adaptive optics" system—unique in 1990—also sharpened and concentrated the light of any circumstellar objects that showed up on the detector.

Since the instrument was meant to prove a concept, Durrance put it together with a collection of lenses ordered from catalogs, a camera design borrowed from Palomar Observatory, and a CCD castoff from Jim Westphal's and Jim Gunn's Hubble Space Telescope instrument. With the adaptive optics coronagraph, dubbed the AOC, exposures were lengthened by factors of a hundred, allowing much fainter objects to be recorded.

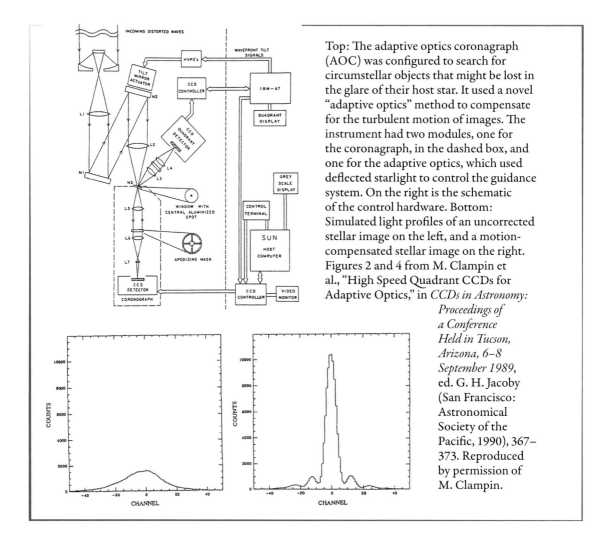

Top: The adaptive optics coronagraph (AOC) was configured to search for circumstellar objects that might be lost in the glare of their host star. It used a novel "adaptive optics" method to compensate for the turbulent motion of images. The instrument had two modules, one for the coronagraph, in the dashed box, and one for the adaptive optics, which used deflected starlight to control the guidance system. On the right is the schematic of the control hardware. Bottom: Simulated light profiles of an uncorrected stellar image on the left, and a motion-compensated stellar image on the right. Figures 2 and 4 from M. Clampin et al., "High Speed Quadrant CCDs for Adaptive Optics," in *CCDs in Astronomy: Proceedings of a Conference Held in Tucson, Arizona, 6–8 September 1989*, ed. G. H. Jacoby (San Francisco: Astronomical Society of the Pacific, 1990), 367–373. Reproduced by permission of M. Clampin.

The AOC's successful maiden run at Las Campanas Observatory caught the attention of Tadashi Nakajima, a Caltech postdoc who became the newest member of the dark-horse team. Having mastered both theory and observations, he astutely recognized the value of collaborating with the Johns Hopkins group. He convinced his colleagues, including his sponsor Shri Kulkarni, that the combination of the 60-inch telescope and the AOC would be as good as or better than any other past or present program at detecting brown dwarfs. In 1992, Nakajima arranged for Durrance and Golimowski to bring their instrument to Palomar, install it at the Cassegrain focus of the 60-inch telescope, and begin a search for brown dwarf companions to stars.[243]

The team's observing program, dubbed the Palomar Billion-Year Survey, was simple, direct, and unambiguous. First, select a couple hundred stars closer than 15 parsecs (around 50 light-years) from Earth. Confirm that the stars have either small space motions or active coronae, which are attributes of stellar youth and an age less than a billion years. Second, image the space surrounding each star in three colors with the AOC. Finally, if a faint red object is detected close to a star, decide whether the two are physically paired or merely aligned by chance. To find out, the team would need to follow the objects for a year to see whether they moved together against the background canopy of relatively stationary stars. If they moved together, it meant that the star and the red object are at the same distance from Earth and physically bound to each other. Since the stars were selected from the Gliese catalog, which lists the precise distances of all nearby stars, the team could then calculate the intrinsic luminosity of the star and the faint red object. If the object was fainter than a normal hydrogen-fusing star, it was likely a brown dwarf or a planet.

The prospect of joining such a promising program to search for brown dwarfs and circumstellar material convinced Ben R. Oppenheimer (now Rebecca Oppenheimer), a physics undergraduate at Columbia University, to enter Caltech for graduate work in 1994, with Kulkarni as sponsor. For her thesis, Oppenheimer—who had long been fascinated with the concepts of high-resolution imaging and adaptive optics—planned to harness the 60-inch telescope for 30 nights a year for five years and bash through the list of target stars.

The lion's share of observing for the project was done by Golimowski and Oppenheimer. The rather hands-on instrument required the pair to repeatedly enter the dark dome during the night to set up each exposure. On their second, long observing run in late October 1994, the two fought exhaustion as they took data for each star on their list. For what would later turn out to be a breakthrough

Chapter 12

> "The little tip-tilt mirror kept the star dead-center while the telescope was rattling around with wind and turbulence, in turn moving the star around. It can look pretty if you're not observing, as in 'twinkle twinkle little star.'"
>
> (R. Oppenheimer, personal interviews with the author, 2008–2016)

observation, Oppenheimer sleepily wrote in the logbook that she saw a faint, red, compact blip next to a target star. The star was designated number 229 in the Gliese catalog and only 5.7 parsecs from the sun. Looking at the data later, she did a double take, blurting out, "Oh my god, what the hell is *this*?" She immediately realized that *this* was something exceptional. The object did not show up at all in the visible-light filter, and barely in the redder-light filters. The two students began leaping with delight around the dome at the thought of what the very red little blip might be. Then reason prevailed as Kulkarni admonished them by phone that the entire team would have to keep their lips sealed until they could reobserve the blip the following year. They had to find out whether the little red blip was, in fact, in orbit around the star.

Hitting Pay Dirt with a Common Little M Star

After eleven months of agonizing silence, on September 14 and 15, 1995, Oppenheimer and Kulkarni were back at Palomar, but this time at the Cassegrain focus of the 200-inch Hale telescope. Since the AOC could not do spectroscopy, they were joined by instrument builder Keith Matthews, who brought one of his specialty instruments—a combination of infrared camera and low-resolution spectrograph with a twist. Since the ratio of brightness between the star and the little red blip was nearly 1,000,000:1, Matthews had attached a metal "finger" that blocked the glare of the central star much like a coronagraph. Happily, the images from the camera showed that the star and blip indeed moved together across the sky. They were gravitationally bound, orbiting companions.

ASTRONOMICAL EXOTICA: NEW FRONTIERS

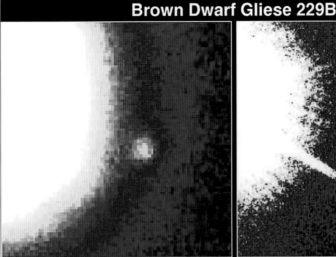

Brown Dwarf Gliese 229B

Palomar Observatory
Discovery Image
October 27, 1994

Hubble Space Telescope
Wide Field Planetary Camera 2
November 17, 1995

The discovery image of the blip is shown in the false-color image, left, taken with the Palomar 60-inch telescope. The blip is less than 8 arcseconds in the southeastern direction from Gliese 229, which is 44 AU or about the distance from the sun to Pluto. The Hubble Space Telescope image, right, was taken high above the turbulence of Earth's atmosphere, where no adaptive optics system is needed. Credit: T. Nakajima and S. Kulkarni (Caltech), S. Durrance and D. Golimowski (JHU), NASA.

As the first raw spectrum of the blip, now officially named Gliese 229B, popped up on the display, Matthews—a veteran observer—blurted out, "Holy shit! There's methane in that thing!"[244] A large chunk of light looked like it had been half eaten away in the wavelength region where methane is known to absorb light. This meant that Gliese 229B had to be cooler than around 1,400 K, since higher temperatures destroy methane's molecular bonds. For comparison, the team took a reflectance spectrum of methane-laden Jupiter, and its features were hauntingly similar to the spectrum of Gliese 229B. Although it is common in biological systems on Earth and in our atmosphere, methane had never before been seen in an object beyond the solar system.

CHAPTER 12

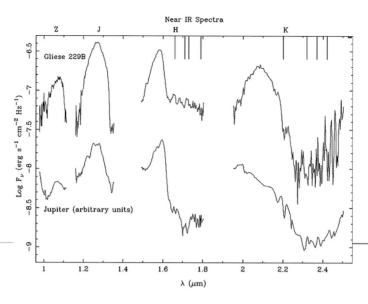

The broadband near-infrared spectrum of Gl 229B, a brown dwarf, was unlike that of any known star, or, in fact, any known celestial object. As a first stab at understanding the object, its spectrum was compared to a spectrum of Jupiter. Although both objects contain methane, Jupiter is a frigid 120 K and very different from Gl 229B, which is estimated to be 1,000 K. The locations of absorption bands due to methane are marked by the vertical bars along the top of the plot in the H and K band passes centered at 1.6 and 2.2 microns, respectively. The spectra of Jupiter are shifted vertically for ease of comparison. Figure 5 from B. R. Oppenheimer et al., "Infrared Spectrum of the Cool Brown Dwarf Gl 229B," *Science* 270 (1995): 1478–1479. Reproduced by permission of Rebecca Oppenheimer.

Takashi Tsuji, a well-known stellar theorist from the University of Tokyo, had in 1964 predicted that unexpected phenomena can appear in the low-temperature/high-pressure conditions prevailing in substellar objects.[245] He explained that the molecules that form in their atmospheres block certain bands of red light from leaving the surface of the star. Thus, the emergent spectrum and color of a cool substellar object is sculpted by strongly absorbing molecules, such as water and methane, that form abundantly in its atmosphere. So—surprise—it turns out that when brown dwarfs cool to a certain temperature, they turn *blue*. This was the opposite of astronomers' reasoning that cooler and redder go together. Unfortunately, Tsuji's 1964 paper had little impact on the astronomical community: nobody searching for brown dwarfs incorporated his results into their search parameters. By looking too far into the red part of the spectrum and with inadequate detectors, the astronomers had been as doomed as the Greek Sisyphus to repeated disappointments.

The Palomar team also benefited from an element of luck. The sensitivity of the adaptive optics coronagraph happened to be a good match to the wavelength regions between bands of strong molecular absorption in Gl 229B's atmosphere. When members of the team returned with the AOC to the 60-inch telescope in October 1995, they found that Gl 229B is 100,000 times fainter than the sun. This is only one-tenth the luminosity of the smallest hydrogen-burning star on the main sequence. Since brown dwarfs younger than 1 billion years are less luminous than that, they should be easily distinguishable from low-mass stars. Together, the temperature, luminosity, spectral signatures, and mass appeared to fit the theoretical predictions for brown dwarfs. Finally, with real data from brown dwarfs, the detailed modeling of their atmospheres could proceed.

> "There are many of us who had looked at Gliese 229 in the past because it is the standard for class M, subclass 2, stars in the Morgan and Keenan stellar classification system, which has been in use since the turn of the 20th century. Many astronomers (perhaps thousands?) had pointed their telescope at Gliese 229. Little did we know that, for all the observations we had done, a poor little brown dwarf was lost in the glare of that host star. When the paper came out, I thought, 'Are you kidding me?' Finding Gl 229B was a *eureka* moment—a watershed moment—in brown dwarf science."
> (D. Kirkpatrick, personal interview with the author, 2015)

Coincidentally, Tsuji, who for three decades had studied and modeled the major observable properties of hypothesized brown dwarfs, published his results in 1995 to enlighten observational astronomers.[246] It was a somewhat precarious venture since the observed properties of M dwarfs were not yet well understood. For realism, Tsuji's models included the same amounts of metals found in the halo and disk of our Galaxy, and the modeled temperatures ranged as low as 1,000 K.

Tsuji and the Palomar team were unaware of each other's work until after the discovery of Gl 229B was announced. Yet the resemblance of Tsuji's models to Gl 229B was a slam dunk: the dark-horse team had the first cool brown dwarf in hand, and Tsuji considered it to be "the most important breakthrough in our exploration of the substellar regime."[247] The discovery was announced in two papers. The first, "Discovery of a Cool Brown Dwarf," appeared in *Nature* on November 30, 1995, first-authored by

Nakajima.[248] One day later, on December 1, "The Infrared Spectrum of the Cool Brown Dwarf Gl 229B" was published in *Science*, with Oppenheimer as lead author.[249]

Although the team continued to image stars on their list, they never found another brown dwarf companion or little red blip. It turns out that brown dwarf companions are much rarer than free-floating brown dwarfs. However, they continued to study Gl 229B with Palomar and Keck telescopes, and shared their data with theorists eager to understand the new class of objects. It had taken three decades of perseverance and significant technological advances to progress from Kumar's prediction of brown dwarfs in 1963 to the first unambiguous discovery in 1995. Brown dwarfs had been long predicted, long awaited, long doubted. But found at last were the progeny of the gas clumps that are part of the continuous spectrum of low- to high-mass celestial bodies formed by nature.

Floating Dirt, Iron Rain, Magenta Skies, and Violent Storms

After 1995, searches for brown dwarfs continued, now based on what an actual brown dwarf looks like instead of having to rely on speculation. Even so, as more objects were identified, Gl 229B stuck out like a sore thumb. The cool temperatures of brown dwarfs created distinctly unstarlike spectra that did not fit the traditional classification sequence for stars. As infrared detectors teased out cooler and cooler bodies, astronomers were forced to incorporate longer wavelengths into their narrow optical-based classification scheme. Davy Kirkpatrick, a research scientist at the Infrared Processing and Analysis Center at Caltech, led the charge to create two new classes: L dwarf, designating warmer objects, and T dwarf—with one methane-bearing member (Gliese 229B)—designating cooler objects. Kirkpatrick thus extended the stellar classification sequence O B A F G K M (with the politically incorrect mnemonic "Oh Be A Fine Girl Kiss Me") that had been established in the early 20th century—the first additions to the sequence after a drought of 80 years.

To Adam Burgasser, a Caltech graduate student, the challenge of finding more methane-bearing brown dwarfs similar to Gl 229B was a scientific carrot dangling in front of his thesis-hungry eyes.[250] Working with his thesis advisors, Kirkpatrick and planetary astronomer Mike Brown, Burgasser used data from the then-newly-released Two Micron All-Sky Survey. Dubbed 2MASS, it had mapped the mass distribution of the infrared universe out to 220 megaparsecs. Fortuitously, the wavelength sensitivity of the filters used by 2MASS perfectly sampled the brightest emission regions of Gl 229B, as the now-famous T dwarf emits the bulk of its radiation from 1 to 2.5 microns.

In two years of intense effort, Burgasser applied successive techniques to ferret out contaminating light sources. First, he applied color and brightness criteria to whittle the 100 million sources contained in 2MASS to 65,000 candidates. Since T dwarfs do not emit light at visible wavelengths, he visually inspected each of the 65,000 candidates to eliminate those that appear on the digitized POSS fields and other optical surveys. He and a night assistant then imaged the 1,000 remaining candidates in the near-infrared with the Palomar 60-inch telescope, banging out two-minute exposures for 10 hours a night for a total of about 60 nights. This eliminated minor planets and spurious sources such as uncataloged asteroids, leaving 200 viable candidates. For the final stage, spectroscopic observations of each object decided whether they were proper T-class brown dwarfs. Out of an initial 100 million objects, the stringent winnowing left Burgasser with only 18 candidates with characteristics similar to the prototype Gl 229B.

Arranging all of the known T dwarfs in a sequence of spectral properties revealed differences in the strength of their features from object to object. Elements such as rubidium, cesium, potassium, iron, and sodium were observed in all of the T dwarfs. The presence of sodium, which absorbs yellow light, meant that "brown" dwarfs might be a misnomer, as they would actually appear magenta to the human eye. Burgasser's search uncovered not only close analogs of Gl 229B but also a broader class of T dwarfs, for which he led the definition of a new classification scheme.[251] But the real jackpot still awaited: besides becoming bluer, the brightness and color of these objects varied dramatically across the transition from L to T classes. Could bands, clouds, and storms in the atmospheres of brown dwarfs be causing those variations?

Brown dwarfs of types L, T, and Y are part of a continuum of sources with atmospheric temperatures that range from thousands of degrees Kelvin to colder than Earth's North Pole. This happens also to be an evolutionary sequence, since brown dwarfs cool over time. Y dwarfs were recently discovered by the Wide-field Infrared Survey Explorer (WISE) satellite—2MASS's successor launched by NASA in 2009. They are the coolest brown dwarfs found so far, and they radiate at even longer infrared wavelengths. Because Earth is awash with a background of 300 K radiation from its surface and atmosphere warmed by energy from the sun, cosmic sources in that temperature range must be studied with satellites. By now thousands of brown dwarfs of various types have been found, some in young clusters, some as companions to nearby stars, and most as faint isolated systems. Because of their faintness, so far all were discovered within a few hundred parsecs of the sun.

Chapter 12

As astronomers began detailed studies and characterizations of brown dwarfs, they found that L dwarfs have thick clouds made of different kinds of metals, minerals, and salts, humorously referred to by some as "floating piles of dirt." As such clouds form, move, and evaporate—much as clouds do on Earth—they produce small variations in the overall brightness of the brown dwarfs. These visible breaks or holes in the cloud cover are reminiscent of Jim Westphal's first images of Jupiter during the days of pioneering infrared detectors at Palomar. In 1969, while scanning across Jupiter's face with a detector sensitive to infrared radiation at 5 microns, Westphal discovered regions of elevated temperature, some the size of North America. Named "hotspots," they were interpreted as localized clearings in the cloud cover that allow interior heat to gush out. Some of the ideas we now have about brown dwarf meteorology were spawned by Westphal's images, taken nearly half a century ago at the 200-inch telescope.

An M dwarf has a clear and cloudless atmosphere because its chemical elements, even iron, are in gaseous form above about 3,000 K. Brown dwarfs become cooler than this. As the temperature drops, the first particles to condense out of the gas are rocks and metals. The condensation—plus a hot, dynamic atmosphere—produces weather in the form of clouds and storms, which are carried along as the object rotates. Instead of the water clouds on Earth or the ammonia clouds on other planets, brown dwarfs have clouds of hot, molten rock floating around in their atmospheres. Imagine a monsoon in the Arizona desert where, instead of water, it rains metals, crystals, or gems.

Although brown dwarfs were at the trailhead of the search to identify dark matter, they are too rare to contribute significantly to the dark mass in the universe. Today we know there are almost as many brown dwarfs as stars in the solar neighborhood. With their wild weather systems, new physics, and exotic chemistry, they have turned out to be fascinating objects on their own. They have helped us understand that the contents of the universe are more diverse than we had previously imagined.

For example, since ancient times, stars and planets had been considered artificially distinct types of objects. That paradigm shifted around the turn of the 21st century. Stars and planets are now thought to be part of a continuum of mass, size, and origin, with the gap between them connected by brown dwarfs. The warmest brown dwarfs join the continuum just below the hydrogen-burning limit of stars, and the coolest merge into the realm of the planets. By coincidence, two precursors of the paradigm shift were announced on the same day (October 6, 1995), at the same location (the "Cool Stars IX Meeting" in Florence, Italy), and at nearly the same hour. During the meeting, the dark-horse team flashed their spectra of Gliese 229B, the first brown dwarf, while Swiss astronomers Michel Mayor and Didier Queloz displayed their spectra of 51 Pegasi, a star whose rhythmic Doppler shift revealed the first planet tugging at an alien sun.

ASTRONOMICAL EXOTICA: NEW FRONTIERS

IN SEARCH OF PALE BLUE DOTS

On Valentine's Day in 1990, NASA's Voyager 1 space probe was headed toward the fringes of our solar system. At the urging of astronomer and author Carl Sagan, it turned its vidicon camera back toward Earth and snapped a portrait of our planetary system, the home where all of human history is encapsulated. From Neptune's orbit 6 billion km away, Earth's light exposed less than one-tenth of one pixel in a frame of 640,000 pixels. That heartbreakingly tiny fleck of light buried in the sun's glare was anointed "pale blue dot." Although the Voyager 1 reconnaissance of the solar system and beyond was highly successful, it taught us a hard lesson. Will humans currently alive ever see an image of an Earth-like planet and know that it harbors life? The answer: likely in your dreams only. The "pale blue dot" was imaged from the edge of the solar system, and typical exoplanetary systems are significantly further away. For example, Proxima Centauri, the nearest star, is 6,600 times further away from Earth than was Voyager 1 when it snapped the famous image. At Proxima Centauri's distance, Earth's light would be almost 44 million times fainter—and completely submerged in the sun's glare.

However, things are not as dire as they may seem. The "pale blue dot" was imaged with only marginal sensitivity, 1970s technology, and a tiny 7-inch aperture—a far cry from 21st-century innovation. A generation ago, we had evidence for only nine planets in the universe (including now-demoted Pluto). Since then, we have made substantial progress: while technology was catching up to viewing extrasolar planets directly, many were already being discovered indirectly. The tiny clockwork fluctuations they impart to their host star's spectrum, or the periodic dimming of their host star's light as the planet transits across the line of sight to Earth, also reveal extrasolar planets. In 2009, NASA launched the Kepler satellite, which spent four years staring at a single patch of sky containing 150,000 stars. Kepler continuously monitored the stars' light, looking for signs of transiting planets. By now, hundreds of thousands of stars are being examined for such rhythmic signatures. More than 2,000 bona fide exoplanets have been cataloged from ground and space observations, and another 4,000 candidates await confirmation.

Initial observations brought many surprises, especially "hot Jupiters," which are huge gaseous planets that orbit frighteningly close to their host star. Because these massive planets produce strong and rapid gravitational tugs, their rhythmic signatures are easy to recognize among hundreds of thousands of candidates. For that reason, indirect methods of searching for exoplanets are biased toward bright, close-in, massive gas giants, instead of Mars-, Earth-, or Venus-type planets. If humans, armed with our current technology, journeyed a few tens of light-years from our own solar system, we would only detect two of our eight planets. The inner rocky planets would be too faint, and Uranus and Neptune orbit the sun so slowly that their rhythms would take years to be recognizable in the data.

The holy grail of planetary science is to take a direct image of an Earth-like planet, resolve individual features on its surface, and capture its spectrum. Hard-core dreamers hope to recognize the chemical fingerprints of some form of life on a world outside the solar system. However, if the brown dwarf Gliese 229B—at a millionth of the brightness of its host star—was hard to image, such dim dots would be much, much harder. The contrast in brightness between a planet and the star it orbits can range from 1:1,000 for hot giant planets to 1:10,000,000,000 for Earth-like planets. This means that an observer would have to block all but one ten-billionth of the light of the star to detect an orbiting Earth-like planet. The highest ratio that ground-based telescopes could achieve until very recently was about 1:1,000,000, short by a factor of 10,000. Furthermore, decoupling the light of a host star from its planetary system requires extremely high resolution. At a distance of 10 parsecs, Earth and the sun would appear to be separated by a mere tenth of an arcsecond. (Recall that an arcsecond is 1/3600 of a degree and the resolution of the human eye is about 60 arcseconds.) How do we cope with these meager odds?

To peel away several layers of difficulty, astronomers are smart about picking host star candidates, and the spectral region in which they observe them. They choose stars that are young, bright, and ablaze with excess amounts of infrared radiation, having figured out long ago that dusty disks around newbie stars are potential hotbeds of planet formation. But the dust doesn't linger. It quickly evaporates as the forming star heats up. Therefore, a visible girdling disk of dust is likely to harbor a freshly formed planetary system still red hot and glowing from the oven. Such planets not only reflect their host star's light but are themselves also luminous, amplifying the number of photons that reach our telescopes. A young star pumps out most of its light in the ultraviolet and blue regions of the spectrum. It is relatively dark in the infrared. On the other hand, planets are brightest in the infrared. This means that the infrared is the best wavelength region to search for planets.

The prototype of such a system turned out to be a young, A-type main-sequence star named HR 8799, 128 light-years from Earth. It displayed the requisite excess of infrared radiation and a dusty disk. In 2010, high-contrast imaging with the 10-meter Keck and 8-meter Gemini telescopes revealed four faint companions orbiting the star, along with two rings of debris.[252] The outer debris ring mirrors the location of our solar system's Kuiper belt and the inner one is analogous to our asteroid belt. Astronomers across the globe rushed to study the system with various highly sensitive state-of-the-art instruments and techniques, and to test whether even more planets or other features would be revealed. Disappointingly, the end result was a proliferation of more images and limited bits of spectra. What happened next, however, is one hell of a story.

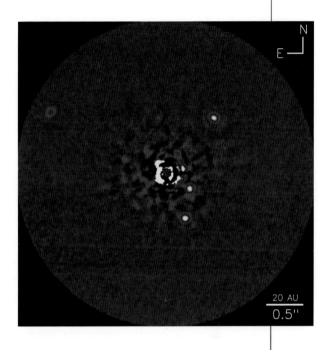

The first image of a planetary system orbiting an alien star bears some resemblance to our solar system. The disk of HR 8799, a massive young star, contains debris fields composed of small rocky, icy objects and tiny dust particles organized into two spatially distinct components. However, the host star produces 1,000 times more ultraviolet radiation than our sun, and the planets detected so far—all gas giants—are five to seven times more massive than Jupiter. The planets orbit counterclockwise. Figure 1c from C. Marois, B. Zuckerman, Q. M. Konopacky, B. Macintosh, and T. Barman, "Images of a Fourth Planet Orbiting HR 8799," *Nature* 468 (2010): 1080–1083, reused by permission of the National Research Council of Canada, C. Marois, and Keck Observatory.

PATHFINDER

It's one thing to see light dots dancing around a star, quite another to know what kinds of atmospheres they have. A spectrum can pick up more than atmospheric features on a distant planet; it can detect signs of life—for example chlorophyll, oxygen, or methane. Surprisingly, the first good chunks of infrared spectra—and images of all four of the planets, all acquired simultaneously—came not from 8- to 10-meter telescopes, but from an intricate chain of new instruments installed in 2012 on the then-64-year-old 200-inch Hale telescope. This feat of technical virtuosity was, in large part, due to the drive and dedication of instrumentalist Rebecca Oppenheimer (previously Ben R. Oppenheimer), who had codiscovered the first cool brown dwarf using the old Johns Hopkins adaptive optics coronagraph (AOC). Although the AOC was successful at imaging gas and dust around very young stars, it was not designed to do spectroscopy, and Oppenheimer now set her goal far beyond mere images of exoplanets. She was determined to include spectroscopy, so that she could model the physical and chemical processes occurring in these objects.

The members of the original brown dwarf team had long since dispersed. Astronaut Sam Durrance, who had helped develop the AOC, had moved to the Institute of Technology in Florida to teach and do research; Tadashi Nakajima received a prestigious award and became physics professor at Palomar College, where he continues his brown dwarf research (with Tsuji); and Shri Kulkarni moved on to other interests. Now chair of the Astrophysics Department at the American Museum of Natural History, Oppenheimer pulled together a team of fresh talent from her own institution, as well as Cambridge University, the Jet Propulsion Laboratory, and Caltech. To severely limit the amount of unwanted starlight reaching the detector, the team had to overcome enormous instrumental challenges. For this reason, it was interdisciplinary, with engineers far outnumbering astronomers. Having an aversion to stilted and forced acronyms, Oppenheimer named the planned chain of high-tech instrumentation and software "Project 1640," since 1,640 nanometers (1.64 microns, or 16,400 angstroms) would be the optimal wavelength of the observations. It took a decade to raise the funds, hash out designs, coordinate the construction, and refine and tune Project 1640 before it was commissioned on the 200-inch telescope.

The reality of observing in the harsh, imperfect environment of a telescope is very different from testing new concepts in a controlled laboratory. Feeling that HR 8799 would be an interesting benchmark for testing the sensitivity and capability of their instruments, the team convened in the data room of the 200-inch telescope in June of 2012. Their long-term goal was to search the vicinity of 200 nearby stars in order to explore the range of planetary characteristics. Along the way, they also hoped to better understand the origin of Earth and to look for signs of life on other worlds.

Because exoplanets are hidden in the glare of their host stars, imaging them with Project 1640 requires a suite of highly specialized instruments and software. Although software controls most of the observations, a team of astrophysicists, instrument specialists, technicians, and computer software experts are present during observations to provide input on targets and to troubleshoot. It is an exhilarating experience to watch the images resembling fireworks as they develop on computer monitors: although only a tiny 4-arcsecond-by-4-arcsecond piece of sky is imaged around each specific target, there's a chance that it will reveal an exoplanetary system.

As the night assistant set the telescope to the coordinates for HR 8799, a sophisticated suite of custom software, stored on 29 computers and 7 electronics racks, took control of the observing sequence. Before the team's eyes, the incoming stellar wave front, fuzzy and wobbly from turbulence in Earth's atmosphere, was quickly sharpened by a powerful adaptive optics system. More than 3,000 actuators deformed a mirror 2,000 times per second to counter the wobbles. The observers then manually moved the star's sharpened image until it was occulted by the disk in the coronagraph. A special wave-front sensor analyzed the speckle pattern introduced by defects in the optics, then smoothed it to help the planets stand out from the background. A multispectral imager simultaneously took images at 32 different near-infrared wavelengths ranging from 10,000 Å to around 18,000 Å. The spectrograph was the clever thesis project of Sasha Hinkley, Oppenheimer's graduate student at the time. With it, Hinkley packed a low-dispersion spectrum into *every single pixel* in the four-megapixel array. That's 40,000 miniature spectra.

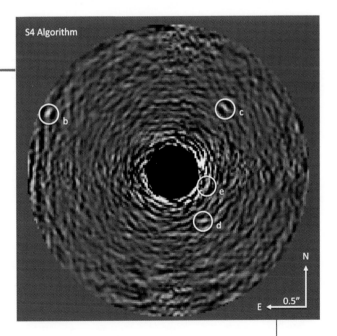

Simultaneous images and spectra of four exoplanets orbiting the star HR 8799 are marked by four circles. The light of the host star, at the center of the blackened circle, has been suppressed with the Project 1640 coronagraph and sophisticated data processing. A pattern of residual speckles, caused by small optical imperfections in the telescope and instrument, litters the background. Buried in each pixel in this image is a miniature spectrum. Once bona fide companions are identified, their spectra and exact positions are extracted for interpretation with complex data analysis software. Image: B. R. Oppenheimer et al., "Reconnaissance of the HR 8799 Exosolar System. I. Near-infrared Spectroscopy," *Astrophysical Journal* 768 (2013): 24, © AAS. Reproduced with permission.

So what is known so far about the planets orbiting HR 8799? The broad molecular features in their spectra do not appear to correspond to any known astrophysical objects. However, some aspects of the planet spectra are similar to those of L- and T-type brown dwarfs, to the night-side spectrum of Saturn, and to models which include clouds. The effective temperatures of the planets orbiting HR 8799 are thought to be around 1,000 K, the same as Gliese 229B, yet the planets have few features in common with Gl 229B or other brown dwarfs. HR 8799, the young, hot, type A5 host star, irradiates its planets with an enormous amount of ultraviolet flux, which may trigger the formation of unfamiliar soots and compounds. Thus, it may not be possible to interpret the spectra of these planets solely on the basis of their mass, age, and metallicity. Published on May 1, 2013 in the *Astrophysical Journal*,[253] the cutting-edge nature of the team's data, methodology, and interpretation was recognized in the astronomical community.

What started as a hunt for dark matter evolved into the hunt for exoplanets, which is becoming more fruitful each year. New technologies have unlocked ever fainter and more nuanced signals. As is so often the case, the strides made by one team toward one objective became the first steps in a relay race that continued to unforeseen destinations.

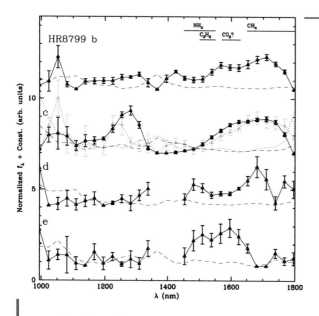

While a planet's orbit and mass can be derived from its location and brightness, its physical and chemical characteristics, such as temperature, composition, and surface gravity, can only be inferred from a spectrum. Shown here is the set of spectra, taken with Project 1640, of four exoplanets orbiting HR 8799. The plot follows the brightness of each object (vertical axis) as the wavelength of observation ranges from 10,000 angstroms to 18,000 angstroms (horizontal axis). The peaks reveal the presence of chemical substances that are part of the planets' atmospheres, and the dips reveal their absence. Except for signs of water, the planets' atmospheres, which potentially contain acetylene, methane, carbon dioxide, and ammonia, are too toxic and hot to sustain life as we know it. Figure 4 from B. R. Oppenheimer et al., "Reconnaissance of the HR 8799 Exosolar System. I. Near-infrared Spectroscopy," *Astrophysical Journal* 768 (2013): 24, © AAS. Reproduced with permission.

Transients, Superluminous Supernovae, and the Future of Observational Astronomy

Whereas in 1995 only a handful of astronomers and planetary scientists were working on exoplanets, today large multidisciplinary teams are inventing some of the most creative and complex devices ever conceived to understand the architecture of exoplanetary systems. Contrary to expectations, we have not yet found another planetary system comparable to ours. For one thing, in our system we don't have any super-Jupiters grazing the sun and being broiled on one side. For another, all of our solar system's planets are in coplanar, nearly circular orbits with perplexingly regular spacings.

What will we learn about the formation of our own solar system from observations of other planetary systems? Will they show us how the Oort cloud arose, or the asteroid belt, or where Pluto came from? We observe extrasolar planetary systems in various stages of formation, with protoplanetary disks and gaps made by shepherding planets clearing away the disk material. Such observations will hopefully allow us to piece together some understanding of these processes in our own solar system.

A far-infrared Spitzer Space Telescope image of the region surrounding HR 8799 may help us understand how planetary systems evolve and interact with their surroundings. This system is younger than 100 million years—compared to our sun's 4.6 billion years—and is still active. Collisions between fragments and small comet-like bodies have created a giant dust-filled halo, shown in yellow and orange. The halo is extraordinarily large—around 2,000 astronomical units across (2,000 times the distance between Earth and the sun). Our entire solar system, including Pluto, would appear as a small dot at its center. The characteristics of the halo may signify a high level of gravitational roughhousing by the massive planets as they settle into stable orbits. Such activities bear a resemblance to the processes that occurred during the formation of our solar system's Kuiper belt. When our solar system was young, Jupiter and Saturn migrated and occasionally sent comets crashing into Earth. The most extreme part of this phase, called "the late heavy bombardment," could explain how our planet attained water. Some comets, acting like wet snowballs, may have delivered water to Earth as they collided with it. Credit: NASA/JPL-Caltech/University of Arizona.

CHAPTER 12

Observing with Project 1640 and other highly specialized and complex instruments means spending years programming computers to assist with data collection and analysis. In addition, during observations, a large team of astrophysicists and technicians are present in the data room to tweak and troubleshoot, and to make real-time decisions about targets. But what if computer-controlled telescopes could acquire images of the sky, and computer algorithms could analyze the data? What if computers could discover and classify cosmic sources, draw inferences, and even trigger follow-up observations on their own? The potential for disruption and surprise is as great as in other areas of life being transformed by artificial intelligence and machine learning.

Such point-and-shoot concepts were taken to the extreme with a daring new kind of robotic all-sky survey inaugurated in 2009 by astronomers, programmers, instrument scientists, and engineers—some permanently stationed on Palomar. The 48-inch wide-field Schmidt camera and the 60-inch telescope were upgraded, automated, and linked into an integrated state-of-the-art system of hard- and software dubbed the Palomar Transient Factory.

The Transient Universe

Transients are spectacular one-stop wonders of the sky that suddenly and unpredictably light up, then fade away over periods ranging from minutes to years. They signal phenomena such as streaking comets, stars being disrupted by supermassive black holes, poorly understood stages of stellar evolution, merging binary neutron stars, and explosions marking the final moments of a star's life. Though their fireworks may be brief, many transients provide a glimpse of gravitational and thermonuclear processes that are otherwise hidden or rarely caught in action.

The idea of a synoptic search for supernovae has its roots in the trailblazing work by Walter Baade and Fritz Zwicky, begun in 1934. The astronomer and the physicist brought their ideas to life with the 18-inch Schmidt camera, the first functioning telescope on Palomar mountain. By conceptualizing this small telescope as a survey instrument for the uncharted sky, Zwicky was far ahead of his time. With an assistant, he spent thousands of nights meticulously guiding the telescope to obtain deep photographic exposures of the sky. Since supernovae are rare in any one galaxy, he maximized his chances of catching one by concentrating his exposures on clusters of galaxies. Then, after a night's observing, Zwicky would lean over a light table and carefully superpose the newest image of the sky with one taken of the same region on a previous night. Covering one eye with a patch, like a pirate, he peered through a low-magnification microscope, shifting the images slightly so that any telltale increase in a star's brightness—signaling a supernova—would pop out. When he could not tease data from sluggish photographic emulsions, he sketched what he saw through the eyepiece. His method was simple yet powerful, and inaugurated a tradition of periodically surveying the sky for supernovae with Schmidt cameras which continues to this day.

In large spiral galaxies like the Milky Way, supernova explosions occur rarely, a few times per century. But how many might occur unnoticed in dwarf galaxies or in the space between galaxies? Why not survey the sky in toto and without bias? In 2009, 73 years after Zwicky began surveying the sky, an automated search at Palomar used a cleverly refined version of his method to find supernovae, and many other transients, by the thousands.[254] Departing from the monthly surveys of the past, the Palomar Transient Factory, or PTF as it is widely known, harnessed the refurbished 48-inch Schmidt camera to image each spot on the visible sky nearly every night. With its increased coverage, PTF caught transients such as supernovae millions of light-years from Earth, exoplanets circling other stars, near-Earth asteroids potentially threatening our planet, and other unforeseen exotic events. Until the end of the 20th century, only a few hundred supernovae, perhaps a thousand, had been known. By 2019, the number had swelled to several 10,000s.

During PTF's nightly observations, there was no human presence in the domes. Instead, computer software controlled the focusing of the telescope and camera, the changing of filters, the opening of the shutter, and the collection and transmission of new data.[255] Through feedback from a sophisticated target scheduler, a weather station, and a data quality monitor, software also controlled pointing the telescope and rotating the dome. At the focal plane of the Schmidt, a high-resolution 100-megapixel mosaic of CCDs captured images of the sky 8 square degrees at a time. In one night, the survey covered 600 square degrees—the equivalent of 3,000 times the apparent area of the moon.

As the shutter closed after each exposure, the raw data were streamed to a computer at the National Energy Research Scientific Computing Center (NERSC) at Lawrence Livermore Laboratory more than 400 miles north. A program to automatically search for supernovae of Type Ia had been in place there since 1988, developed by one of the teams that discovered the accelerating expansion of the universe. The software had since been upgraded to serve PTF's higher cadence and all-sky blind-search methods. The process of sifting through the data to find potential transients that might merit a second look—work that was formerly assigned to undergraduates—was now completely automated. Sophisticated computer code with lighthearted names such as Scamp and Swarp and HOTPAnTS stood ready to process the new image by subtracting from it, pixel by pixel, an archived reference image of the same part of the sky. Then, like an automated prospector panning for gold, Terapix SExtractor, a machine learning algorithm developed by UC Berkeley professor Josh Bloom and his team, evaluated the output to distinguish real celestial objects from digital artifacts such as cosmic-ray traces and poorly subtracted stars. During the 2008 commissioning of PTF, the computers were trained on baseline images of actual transients prepared by a pool of 30 experts. The two-choice decision-making process was humorously likened to a farmer "sexing the newborn chickens."

If a cosmic light source appeared or disappeared from one night to the next, the event was flagged as a potential transient of scientific interest. The data were then uploaded to computers at the nearby campus of the University of California at Berkeley. Taking into account various observed features of the transient and its environment, programs with intelligent algorithms classified the event as a variable star, a gamma-ray burst, a supernova, etc. Scientifically promising events could automatically trigger Palomar's robotic 60-inch telescope to measure the transient's color and to generate its light curve. Amazingly, all of this automated imaging, pipelining, data processing, decision making, and follow-up would take less than about 40 minutes after the 48-inch Schmidt opened its shutter. Although the supernova would have grown and evolved during those 40 minutes, the response time would still be bullet-fast compared to decades earlier, when the timescale for human processing could be days to weeks. At this point, the decision of whether an event of interest required follow-up observations from other telescopes would have been passed on to the astronomer-on-call, who would be paged by a computer in the middle of the night.

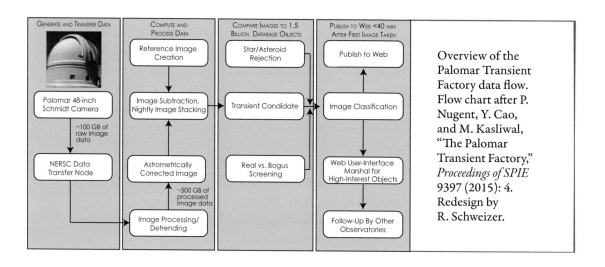

Overview of the Palomar Transient Factory data flow. Flow chart after P. Nugent, Y. Cao, and M. Kasliwal, "The Palomar Transient Factory," *Proceedings of SPIE* 9397 (2015): 4. Redesign by R. Schweizer.

On the night of August 23, 2011, the 48-inch Schmidt camera was pointed in the direction of the Big Dipper's handle. Bingo! A new spot of light was recorded in Messier 101, the Pinwheel galaxy. But this night was different. A software glitch had jammed the system, bringing to a halt the chain of command in the pipeline connecting the 48-inch at Palomar to the computers at NERSC. Hence there was no alert sent by the computer until Peter Nugent, a theoretical astrophysicist at UC Berkeley and head of the real-time classification group at the Lawrence Berkeley National Laboratory, managed to fix the glitch around noon the next day. He was playing catch-up with the previous night's haul when he spotted the newly flagged transient in M101. It appeared too faint to be a supernova yet too bright to be a nova, and no source had been visible at those coordinates the night before. The event, now tagged PTF11kly, seemed worth pursuing. Since crucial data from the first hours after a stellar explosion cannot be recovered later, he urged members of the multi-institutional PTF Consortium to observe the new nova or supernova candidate without delay.

Things became even more exciting during the next few hours, as the 2-meter robotic Liverpool Telescope at La Palma in the Canary Islands took spectra of PTF11kly, and NASA's Swift satellite began taking a series of images at ultraviolet wavelengths. As Nugent scrutinized the incoming spectra, he recognized features of a Type Ia supernova, a standard candle used to measure distances of cosmic objects. Since the supernova was relatively faint intrinsically, it was likely an exceptionally young explosion. Later, follow-up photometry would nail down the age of the supernova—only 11 hours old when it was first recorded by the Palomar 48-inch Schmidt camera. Nugent immediately issued an "Astronomer's Telegram" encouraging astronomers worldwide to observe the supernova. Blazing from a spiral arm of Messier 101—only 21 million light-years away—this was the closest and brightest Type Ia supernova found in nearly four decades. A rare feat and a shining moment for the Palomar Transient Factory, it validated the ideas behind automated high-speed pipelines, machine learning algorithms, blind synoptic surveys, and one-day cadences. And PTF's old but souped-up telescopes took a front-row seat to one of nature's most powerful explosions.

The Pinwheel galaxy is a face-on spiral that is double the size of the Milky Way galaxy and loaded with as many as a trillion stars. Light from the explosion of Supernova 2011fe reached Earth on August 23, 2011, and was recorded by the Palomar 48-inch Schmidt camera about 11 hours after ignition. Credit: B. J. Fulton, Las Cumbres Observatory Global Telescope Network.

Chapter 12

The Fountainhead

For decades astronomers thought that a supernova of Type Ia occurs when a white dwarf in a binary star system syphons gas from its companion. After some time, the star's mass exceeds a tipping point and runaway carbon fusion ignites in its interior. In a heartbeat, a thermonuclear shock wave races out from the core, fusing everything in its path into heavy metals—foremost, radioactive nickel. The dying star's hellish interior is chaotic and turbulent.

Yet, miraculously, in the spectra of SN 2011fe taken less than 28 hours after its detonation, knots of unfused oxygen were seen hurtling out in front of the shock wave at more than 20,000 km/s, with knots of unfused carbon trailing close behind. These elements contained clues to the supernova type.

Supernovae of Type Ia are significant diagnostic tools. Yet, until SN 2011fe was discovered, no recent Type Ia supernova had occurred close enough to Earth to give astronomers the opportunity to learn about—and possibly identify—the elusive progenitors of these explosions. Previous theoretical modeling and data had implicated an exploding white dwarf, yet the nature of its companion star was a lingering puzzle. The detection of oxygen and carbon—but no hydrogen—in SN 2011fe confirmed that at least in this case the exploded star was a so-called "carbon-oxygen" white dwarf. Further confirmation came from a 0.4-meter robotic telescope in Mallorca, which happened to have imaged the Pinwheel galaxy only four hours after the explosion, yet had recorded no sign of the supernova. Comparing this finding to his theoretical models, Caltech postdoc Anthony Piro[256] put tight constraints on the size of the progenitor: it had to be less than 1/50th the size of our sun, which strongly argued for an exploding white dwarf.

Theoretical models had predicted for some time that the white dwarf's companion could be either a red giant star a hundred times more luminous than the sun, a main-sequence star about as luminous as the sun, or another white dwarf about 1,000 to 10,000 times less luminous than the sun. To decide among these three possibilities, Weidong Li—an expert in precise measurements of stellar positions and brightnesses—got involved. Li had risen to his position as an astrophysicist at UC Berkeley from a childhood on a Chinese farm, where he was the first in his village to attend college. Now he found himself a cherished member of Alex Filippenko's research group, scrutinizing images of the Pinwheel galaxy taken by the Hubble Space Telescope *before* the explosion of SN 2011fe. He was trying to decide among possible companion stars whose brightnesses differed by a factor of a million. If a red giant star had been the supernova's companion, it should have been bright enough at the Pinwheel's distance to be recorded by the Hubble, yet there was no such star visible at the location. Li was certain the companion star was not a red giant, but the Hubble images did not go deep enough to detect a fainter main-sequence star like the sun, or a white dwarf. On December 1, 2011, Li and Nugent reported their findings in the journal *Nature* in separate papers.[257]

Fortunately, there was an additional piece of information to help decide. If the companion star was a red giant, models predicted that the material ejected by the exploding white dwarf would create a shock wave when it slammed into the giant, heating it to the flash point within a day or two of the explosion. Yet such a bright flash of light had not been observed. Therefore the companion had to be the size of the sun or smaller. This finding added credence to Li's work. The absence of any hydrogen emission lines in a spectrum of SN 2011fe taken nearly three years later further strengthened the conclusion that the supernova had been produced by two merging white dwarfs,[258] both devoid of hydrogen. The companion was less likely to be a sunlike star, which would have had hydrogen in its atmosphere.

The discovery of a nearby Type Ia supernova shortly after its progenitor's explosion, and the opportunity to study its evolution daily across the spectral range from the ultraviolet to the infrared, was exciting for several reasons. First, only a dozen years earlier, Type Ia supernovae had served as standard candles in the discovery of the accelerating expansion of the universe and the inference of dark energy. They were used empirically, without a detailed understanding of their progenitors and the physics of the explosion. Second, Type Ia supernovae—along with their core-collapse cousins—play multiple roles in universal ecosystems. Both types drive massive outflows of gas, ionizing radiation, and energetic particles that influence the formation of future stars, and the energy balance and structure of galaxies. Third, some supernovae are linked to gamma-ray bursts, and others leave behind exotic objects such as neutron stars or black holes. From observations of such Type Ia supernovae out to great distances, cosmologists can deduce the expansion history of the universe and test various models of dark energy.

Besides being viable standard candles for cosmological studies, supernovae of Type Ia are also the main contributors of iron-peak elements to the chemical abundances in the universe. During their runaway chain of nuclear-fusion reactions, a white dwarf star—formerly an inert globe of carbon and oxygen ashes—transforms itself phoenix-like into a nutrient-rich bomb. At detonation, 1 billion suns' worth of light—and a star's worth of debris rich in synthesized heavy elements—are forcefully injected into the interstellar gas of the host galaxy. As a result, the chemical composition of the universe has evolved over eons, from a simple soup of hydrogen, helium, and lithium to all 92 natural elements present in stars, planets, and life as we know it. The many Type Ia supernovae that exploded before the formation of our solar system about 4.5 billion years ago contributed their synthesized iron and other heavy elements to the solar nebula—and therefore set the stage for the evolution of life on Earth.

Chapter 12

Until recently, supernovae were thought to be of only two spectral types, as laid out by Rudolph Minkowski in 1941: Type I, with broad emission lines and no hydrogen lines, and Type II, with strong Balmer emission lines of hydrogen. Then in 2007, Robert Quimby, a graduate student at the University of Texas, Austin, discovered an anomalously bright supernova that was of neither type, a new brighter kid on the block. Four years later and by then a Caltech postdoc, he and his team discovered four more such very luminous supernovae while scrutinizing data from the Palomar Transient Factory. Spectra taken with the 200-inch telescope, one of the 10-meter Kecks, and the 4.2-meter William Herschel telescope on the Canary Islands showed no hydrogen lines whatsoever. At first, the objects seemed to be extremely bright outliers of Type I supernovae, but unrelated to each other. Then Quimby began to notice similarities between their spectra when he aligned them to the same redshift. Now regarded as members of a new class of objects[259] called "superluminous supernovae," they produce such a large amount of radiation that their peak luminosities far exceed those of normal supernovae, sometimes by a factor of more than 100. As they are visible up to 10 times farther away than typical supernovae and they tend to remain near peak brightness for several weeks, superluminous supernovae are promising cosmological tools.

Although superluminous supernovae occur mostly in metal-poor dwarf galaxies with only a few billion stars, they come from massive progenitors, some being stars of more than 40 solar masses. Given the diversity of their light curves, spectra, and environments, the extreme physics of their explosions and the mechanisms by which they generate energy remain controversial. Rare events such as superluminous supernovae would certainly be missed without modern high-cadence all-sky surveys such as PTF.

Palomar is one of several observatories that have formed international consortia to undertake next-generation surveys that will monitor the sky even more frequently and to greater depth than hitherto possible. The newest version of PTF is the Zwicky Transient Facility (ZTF),[260] named in honor of Fritz Zwicky for his contributions to the study of supernovae. In an obituary, Harvard professor Cecilia Payne-Gaposchkin wrote, "Looking back on his rugged determination and his slightly Renaissance flavor, one is reminded of Tycho Brahe: brilliant, opinionated, combative, a superb observer, and a very human person."[261] Payne-Gaposchkin judged that Zwicky's ideas were so fertile and his projects so vast that he could have employed all the facilities of a great observatory. In operation since 2017, Zwicky's namesake is bigger, faster, more efficient, and more highly automated than was PTF, yet a human is still on call to receive email from an automated "broker." A giant 576-megapixel camera—with sixteen CCDs covering the 47-square-degree focal plane of the 48-inch Schmidt—snaps images of the sky every 30 seconds to a depth of 20.5 magnitudes. In the early 1950s it took a full nine years to photograph the entire northern hemisphere of the sky with the Schmidt. Now it can be reimaged in two nights with ZTF. The new data-processing system—a major upgrade built on lessons learned from the PTF data-processing pipeline—is housed in the Infrared Processing and Analysis Center at Caltech. ZTF is finding treasures such as transiting planets around young stars, superluminous supernovae, evidence for the existence of Planet Nine, and optical counterparts to the gravitational-wave sources detected by LIGO. The project aims to eventually catalog more than a billion objects for further study.

In another stunning discovery based on ZTF data, two white dwarfs, each the size of Earth, were found orbiting each other in only 6.91 minutes.[262] Their orbit is so compact that the entire binary system would fit inside the body of Saturn! Named ZTF J1539+5027 (and nicknamed J1539), this twirling pair of white dwarfs was discovered by Caltech graduate student Kevin Burdge while searching for periodic sources among the 20 million ZTF light curves available at that time. A peculiar pattern of blinking light, the signature of a short-period eclipsing binary star, caught his attention. Burdge then set an international team to work. On the same night, observations with the Kitt Peak National Observatory 84-inch telescope revealed a deep dip in the light curve every 6.91 minutes, confirming the system as an eclipsing binary. Then CHIMERA, a high-speed imaging photometer on the 200-inch telescope, observed the eclipse and revealed a second dip in the light curve due to the secondary star. The primary eclipse occurs when the larger, dimmer star passes in front of the smaller, brighter star—which is 10 times hotter than the sun—and blocks its light every 6.91 minutes. The two eclipses are visible because the orbit is oriented nearly edge-on to the line of sight from Earth. The work of Kevin Burdge and his team was announced on July 25, 2019.

This pair of closely orbiting white dwarfs will spiral together as the system loses energy due to its emitting gravitational waves. The two dwarfs may eventually merge into a single star, or one may be gradually shredded by the other. General relativity predicts that their orbital period should measurably diminish on the time scale of a few years. When the ZTF data were compared with archived PTF data taken over the past ten years, this prediction was confirmed. The time estimated for the two stars to merge is of the order of 200,000 years. Whether they will then explode as a supernova of Type Ia or shred each other to pieces, the grand finale promises to be spectacular.

Chapter 12

⋯

What started in the 1930s as a quest for "dark matter" has brought astronomers to the edge of understanding the brightest explosions in the universe. In a story woven with universal themes of scientific discovery, astronomers seeking to characterize the mysterious dark matter that holds galaxies together developed techniques to look for dim objects in bright places. Instead of finding dark matter, they found brown dwarfs and developed high-resolution technology to hunt for exoplanets. This technology—required to someday find other worlds similar to Earth—has given rise to rapid and deep whole-sky surveys. These surveys are themselves ever improving through better telescopes, detectors, computers, and analytical tools. Recently, whole-sky surveys have revealed not only thousands of exoplanets but also rich populations of once-rare phenomena that allow astrophysicists to probe the origins of the universe, bringing the meandering path of discovery full circle.

Compared to these prodigious new datasets, the data that supported discoveries in the second half of the 20th century may seem primitive. Yet astronomers and engineers tenaciously extracted paradigm shift after paradigm shift from them. These pioneers, too, were enabled by technology, emboldened by curiosity, and open to serendipity. Some guarded their nascent insights with trickery and their observations with subterfuge. At times their rivalries drove innovation and experimentation in the race to discover and publish first. Just as often, though, these pioneers sought complementary expertise, dismantled old conflicts, and helped each other.

I started writing this book with a concept of twelve chapters standing on their own, each picking off one class of discoveries. But I rapidly found that no story of modern astrophysics can be cut into pieces so cleanly. Like many other fields, observational astronomy progresses through new technology, the availability of ever more massive data collections, new analytical methods, and the chance discovery of strange entities or phenomena beyond our current understanding. That is where curiosity, creativity, and technology will continue intersecting with serendipity to expand our understanding of the universe. Future revelations will likely be more complex, yet no less deeply intriguing.

ASTRONOMICAL EXOTICA: NEW FRONTIERS

Discovered among large-survey data, arcs of light from a distant background galaxy nearly encircle the central region of this rich cluster of galaxies, forming a rare and exotic phenomenon known as an Einstein ring. Such arcs are an elegant manifestation of Einstein's theory of general relativity, which predicts that a massive structure warps the geometry of its surrounding spacetime and can act as a gravitational lens. Here, the light from a distant galaxy is being deflected and distorted by SDSS J0146-0929, a massive galaxy cluster with hundreds of members that lies along the line of sight to Earth. On August 23, 2010, the Triple Spectrograph on the Palomar 200-inch telescope confirmed a redshift of $z \sim 2$ for the arcs. Thus, the distant galaxy is about five times further away from us than the foreground lensing cluster. Notice the bluish color of the arcs, which indicates a high rate of ongoing star formation in the distant galaxy. As in a powerful cosmic telescope, gravitational lensing intensifies and magnifies background structures, providing a view of objects that are otherwise too far and too faint to study. The strength of the magnification, the number of arcs, and their overall pattern enable astronomers to derive the *total* mass—both visible and dark—of the intervening cluster, thus providing a window into dark matter. The detection of absorbing gas, ionized gas, and even the internal kinematics of such distant galaxies allows astronomers to determine elemental abundances and the history of star formation in the early universe. This Einstein ring was discovered during a search for arcs in fields imaged during the Sloan Digital Sky Survey. D. P. Stark et al., "The CASSOWARY Spectroscopy Survey: A New Sample of Gravitationally Lensed Galaxies in SDSS," *Monthly Notices of the Royal Astronomical Society* 436 (2013): 1040–1056. Credit: ESA/Hubble and NASA; acknowledgment: Judy Schmidt.

Acknowledgments

Any book project is its own bustling ecosystem. Just as Palomar astronomers sparked each other's insights and shared their work, I have benefited from the support of academics, friends, and family. Many of the people I thank here have become friends, bonding over writing, research, the observatory, and our shared purpose.

I am profoundly grateful to The Rockefeller Foundation for a generous project grant, and I am especially indebted to former President Judith Rodin for a supplemental grant from the President's Discretionary Fund. Both of these grants allowed me to focus on project research and the writing of this book. My program advisors Darren Walker, Donald Roeseke, Jr., and now Emily Prandi have been singularly helpful and encouraging. The Emerge program of Fulcrum Arts, directed by Robert Crouch, provided fiscal sponsorship and joyous connection to a diverse community of creative souls at the intersection of art and science; Blue Trimarchi of Art Works Fine Art Publishing, Inc, digitized and printed the Russell W. Porter drawing for this book; and Caltech Optical Observatories kindly provided a small startup grant that was helpful during the planning stages.

Caltech President Jean-Lou Chameau offered me long-term Visitor status, granting access to Palomar Observatory, a campus office, and visiting researchers from across the globe. I thank Kenneth A. Farley, Chair of the Division of Geological and Planetary Sciences, for his warm hospitality, and the Division of Physics, Mathematics, and Astronomy for office space. Andrew Lange was an inspiration to me in many ways; he was warm and altruistic, and I miss him. The beloved, inimitable Harry B. Gray, with whom I taught science writing, graciously and cheerfully rallied behind this book project. Maarten Schmidt offered especially endearing, unfiltered responses during several interviews. Patrick Shopbell provided much-appreciated computer advice. My dear friends Irma Black, Marcia Brown, Marcia Hudson, and Lindsay Cleary adeptly ironed out administrative wrinkles, and Luwam Habte provided early clerical assistance. Mansi Kasliwal taught me about "sexing the chickens." I am deeply grateful to several astronomers, among them Lee Armus and Rebecca Oppenheimer, for inviting me to participate in their observing runs at Palomar. I thank Christoph Baranec and Reed Riddle, my hardworking Robo-AO subbasement officemates, for informative chats; Peter Mason for interesting tales of his grandfather Max, a former president of The Rockefeller Foundation; JoAnn Boyd for helpful conversations; Ken Farley and

George Djorgovski for kindly temporarily sharing their own offices with me; and Bill McLellan, a 50-year veteran of Palomar, for gleefully showing me Palomar history treasures in the Vault.

Conversations with Richard Preston (author of *First Light*) were riotously fun, immensely informative, and strongly reaffirming to a first-time book author. Maverick James E. Gunn shared stories of his long and extraordinarily productive association with Palomar and the Sloan Survey—all while we explored Pasadena on foot in search of watering holes. Allan Sandage candidly revealed his inner thoughts and emotions as he relived early days at Mount Wilson and Palomar, sometimes while we baked bread competitively. Conversations with Stephen Hawking on the importance of broadcasting science to the public, and our convergence of purpose, inspired and fortified me while teaching science writing at Caltech and, later, working on this book. Kip Thorne, an exemplar of eloquent writing, devoted teaching, and roving intellect, offered insight and advice while being a good sport about revealing Stephen's secret hideouts.

It was electrifying to hear John Grunsfeld, who helped repair and upgrade the Hubble Space Telescope, characterize his observing at Palomar as "a big, romantic adventure," a feeling that carried over into space. Robin and Todd Mason began their captivating TV documentary *Journey to Palomar* before I began this book and were fountains of support and insight. Lisbeth Bosshart Merrill unflinchingly helped clear impediments with many hours of research, discussion, and counsel, sometimes with her associate David G. Rosenbaum—a potent team. Dale Vrabec, a Caltech graduate student in 1948, shared precious memories of Russell W. Porter at his drawing board. On Mount Wilson, where the strong perfume of junipers hammers the senses as thoughts of the telescopes' inroads into the unknown universe invigorate the mind, Samuel Hale and Brack Hale told stories of their grandfather George Ellery Hale— may this book help them appreciate the wonders of his greatest achievement.

Sean Carroll, Kip Thorne, Dava Sobel, Richard Preston, Marcia Bartusiak, Dennis Overbye, Michael Lemonick, Ann Finkbeiner, Ronald Florence, and Donald Goldsmith: thank you all for sharing your experiences in writing, agenting, and publishing to sustain and guide me on the long road to completion. Dava even confided her book-launching mantra to me. I thank Jay Pasachoff for publishing advice; Jeff Dean for perceptive conceptual advice; and Kevin Starr for historical perspective.

I respect and admire the generations of Palomar mountain folks who— tirelessly on-call—kept the telescopes, instruments, and astronomers functioning through the night, the years, and the decades, all while keeping the place spotless and gleaming. Hearty thank-yous to Bruce Baker for an insider's tour of the Phantom, the railroad tracks, and the welded and bolted steel structures, and for showing me how to wash, rinse, dry, and aluminize a 14.5-ton mirror; to Mike Doyle for eschewing the elevator so we could scramble up the narrow ladders to the prime-focus cage; to Steve Einer for demystifying the Rube Goldberg-style vacuum pump, the artful bell jar with its aluminum twisties, and Gertrude the rubber chicken; to the keepers of the Monastery and diPali the gourmet cook; to the legendary Jean Mueller, Kajsa Peffer, and others for virtuoso slewing, setting, and the "we're there" at the Hale telescope; and to Andy Boden and Dan McKenna for facilitating my numerous trips to Palomar. Rick Burrus, Loyd Caster, Bruce and Dana Cuney, Steve Flanders, Carolyn Heffner, John Henning, Scott Kardel, Steve Kunsman, Hal Petrie, Drew Roderick, Hector Rodriguez, Kevin Rykowski, Pam Thompson, Greg van Idsinga, Ernie Velador, and Jeff Zolkower were sweetly tolerant of my questions and my peering over their shoulders. A salute to the dedicated docents of Palomar. I love you all and apologize if I have inadvertently left any of you out.

Several willing souls read and commented on individual chapters, improving accuracy and presentation. I greatly appreciate the time and candor of Eric Becklin, Michael Brown, Hannes Bühler, Adam Burgasser, Alan Dressler, Richard Green, James E. Gunn, Simon Herrmann, Robert Kirshner, Todd Mason, Rebecca Oppenheimer, Anthony Piro, Allan Sandage, Linda Stryker, Noel Swerdlow, Alar Toomre, and Robert Zinn. I owe a large debt of gratitude to Alexei Filippenko and François Schweizer, who independently read and annotated the entire manuscript. Everyone's comments and questions led to revisions that clarified and augmented the science and the stories. Any surviving errors are mine alone.

This book would have been much drier without the personal recollections of many people who were intimately involved with scientific research at Palomar, or with its history. They gave generously of their time, allowed me peeks into their methodology and motivation, and contributed to a realistic portrait of how science is done. I interviewed (mostly in person) Helmut Abt, Kurt Adelberger, Phil Appleton, Lee Armus, Halton (Chip) Arp, Gibor Basri, Konstantin Batygin, Eric Becklin, Chas Beichman, Edo Berger, Alexander Boksenberg, Todd Boroson, Michael Brown, Robert Brucato, Adam Burgasser, Judy Cohen, Marshall Cohen, David DeVorkin, S. George Djorgovski, Alan Dressler, Sam Durrance, Richard Ellis, Alexei Filippenko, Andy Fruchter, Peter Goldreich, Miller Goss, Richard Green, John Grunsfeld, James E. Gunn, Stephen Hawking, Luis C. Ho, John Huchra, Andrew Ingersoll, John A. Johnson, Ken Kellerman, Ivan King, J. Davy Kirkpatrick, Robert Kirshner, Charles T. Kowal, Shri Kulkarni, Andrew E.

Lange, Ken Lawrence, Donald Lynden-Bell, Christopher Martin, Yuichi Matsuda, Keith Matthews, Thomas A. Matthews, Dimitri Mawet, Guido Münch, Bruce Murray, Gerry Neugebauer, Phil Nicholson, Rebecca Oppenheimer, Glenn Orton, Eric Persson, Max Pettini, Mark M. Phillips, Thomas Phillips, Sterl Phinney, Andrew Pickles, Anthony Piro, George Preston, Michael Rauch, Martin Rees, Brant Robertson, Vera Rubin, Allan Sandage, David Sanders, Anneila Sargent, Wallace Sargent, Maarten Schmidt, François Schweizer, Leonard Searle, Eugene Serabyn, Alice Shapley, Stephen Shectman, Chris Shelton, Carolyn Shoemaker, Thomas Soifer, Charles Steidel, Edward Stone, Gustav A. Tammann, Jill Tarter, Richard Terrile, Kip S. Thorne, Thomas Tombrello, Alar Toomre, Charles Townes, Virginia Trimble, Gerald Wasserburg, Mike Werner, and Kevin Xu. Thank you all!

It has been a pleasure to work with the MIT Press. My acquiring editor Jermey Matthews enthusiastically took on this project and worked with me to develop a common vision for the book. The manuscript was then masterfully guided to publication by the editorial, design, production, and marketing teams, among them Matthew Abbate, Yasuyo Iguchi, Susan L. Clark, Katie Hope, Jay McNair, Nicholas DiSabatino, Helen Weldon, and Haley Biermann. Much appreciation for your efforts.

Robert Wayne, Blaire Van Valkenburgh, Jua, Pacholi, and Zuni provided a lovely, isolated writer's escape at Condor Crag.

I am grateful to my parents for letting their young daughter roam free at the Griffith Observatory in Los Angeles, a candyland of images, exhibits, and scientific delights that fueled my love of astronomy. I also thank Director Edwin Krupp for a more recent behind-the-scenes tour. Last but not least, I express my deepest love and gratitude to my family for their tireless support of this project over the years and for plotting to refresh my muse with getaways to exotic places. I greatly appreciated François's keen insights, Briana's dazzling artistry in developing the design, Maia's clever application of "McKinsey analyses," Rena's deft construction of spreadsheets and dogged pursuit of permissions, and Teia's sagacious comments while editing, plus the lending of her cat. My heart is full.

Notes

Preface

i   Kevin Starr, PhD, was an award-winning California historian, author, and professor at the University of Southern California. Personal interviews with the author, 2009.

ii   John Grunsfeld, PhD, walked in space eight times, serviced the Hubble Space Telescope three times, braved riding on 4.5 million pounds of explosive fuel to accomplish his missions, and in 1995 carried a piece of Palomar on a coveted ride into space aboard space shuttle *Endeavor*. Personal interview with the author, 2010.

Chapter 1

1. R. B. Fosdick, *Adventure in Giving: The Story of the General Education Board, a Foundation Established by John D. Rockefeller* (New York: Harper and Row, 1962).
2. L. A. DuBridge, *The Men of Palomar. Dedication of the Palomar Observatory and the Hale Telescope* (Pasadena: California Institute of Technology, 1948), 34.

Chapter 2

3. E. P. Hubble, "Explorations in Space: The Cosmological Program for the Palomar Telescopes," *Proceedings of the American Philosophical Society* 95, no. 5 (1951): 461–470.
4. W. Baade, "The Period-Luminosity Relation of the Cepheids," *Publications of the Astronomical Society of the Pacific* 68 (1956): 5.
5. W. Baade, "A Program of Extragalactic Research for the 200-Inch Hale Telescope," *Publications of the Astronomical Society of the Pacific* 60 (1948): 230.
6. Interview of Henrietta Swope by David Devorkin, August 3, 1977, Niels Bohr Library & Archives, American Institute of Physics, College Park, MD.
7. M. L. Humason, N. U. Mayall, and A. R. Sandage, "Redshifts and Magnitudes of Extragalactic Nebulae," *Astronomical Journal* 61 (1956): 97–162.
8. Baade, "A Program of Extragalactic Research for the 200-inch Hale Telescope," 230.
9. A. Sandage, "Current Problems in the Extragalactic Distance Scale," *Astrophysical Journal* 127 (1958): 513.
10. E. P. Hubble, "The Law of Red Shifts (George Darwin Lecture)," *Monthly Notices of the Royal Astronomical Society* 113 (1953): 658.

11   A. Sandage, "The Ability of the 200-Inch Telescope to Discriminate between Selected World Models," *Astrophysical Journal* 133 (1961): 355.

12   G. A. Tammann and A. Sandage, "The Stellar Content and Distance of the Galaxy NGC 2403 in the M81 Group," *Astrophysical Journal* 151 (1968): 825.

13   A. Sandage, "The First 50 Years at Palomar, 1949–1999: The Early Years of Stellar Evolution, Cosmology, and High-Energy Astrophysics," *Annual Review of Astronomy and Astrophysics* 37 (1999): 445–486.

14   M. Schmidt, "3C 273: A Star-like Object with Large Red-Shift," *Nature* 197 (1963): 1040.

15   G. O. Abell, "The Distribution of Rich Clusters of Galaxies," *Astrophysical Journal Supplement Series* 3 (1958): 211.

16   F. Zwicky, E. Herzog, and P. Wild, *Catalogue of Galaxies and of Clusters of Galaxies*, 6 vols. (Pasadena: California Institute of Technology, 1961–1968).

17   A. Sandage, J. Kristian, and J. A. Westphal, "The Extension of the Hubble Diagram. I. New Redshifts and BVR Photometry of Remote Cluster Galaxies, and an Improved Richness Correction," *Astrophysical Journal* 205 (1976): 688–695.

18   A. Sandage and E. Hardy, "The Redshift-Distance Relation. VII. Absolute Magnitudes of the First Three Ranked Cluster Galaxies as Functions of Cluster Richness and Bautz-Morgan Cluster Type: The Effect on q_0," *Astrophysical Journal* 183 (1973): 743–758.

19   A. Sandage, "The Redshift-Distance Relation. II. The Hubble Diagram and Its Scatter for First-Ranked Cluster Galaxies: A Formal Value for q_0," *Astrophysical Journal* 178 (1972): 1–24.

20   Sandage, "The First 50 Years at Palomar."

21   J. E. Gunn and J. B. Oke, "Spectrophotometry of Faint Cluster Galaxies and the Hubble Diagram—An Approach to Cosmology," *Astrophysical Journal* 195 (1975): 255–268.

22   B. M. Tinsley, "Evolution of the Stars and Gas in Galaxies," *Astrophysical Journal* 151 (1968): 547.

23   J. E. Gunn, personal interviews with the author, 2008–2017.

24   F. Zwicky, "On the Search for Supernovae," *Publications of the Astronomical Society of the Pacific* 50 (1938): 215.

25   W. Baade, "The Absolute Photographic Magnitude of Supernovae," *Astrophysical Journal* 88 (1938): 285.

26   R. Minkowski, "Supernovae and Supernova Remnants," *Annual Review of Astronomy and Astrophysics* 2 (1964): 247.

27   R. Minkowski, "Spectra of Supernovae," *Publications of the Astronomical Society of the Pacific* 53 (1941): 224.

28 C. T. Kowal, "Absolute Magnitudes of Supernovae," *Astronomical Journal* 73 (1968): 1021–1024.
29 R. Kirshner, personal interviews with the author, 2009–2013.
30 R. P. Kirshner et al., "Spectrophotometry of the Supernova in NGC 5253 from 0.33 to 2.2 Microns," *Astrophysical Journal* 180 (1973): L97.
31 J. L. Greenstein and R. Minkowski, "An Atlas of Supernova Spectra," *Astrophysical Journal* 182 (1973): 225–243.
32 I. P. Pskovskii, "Light Curves, Color Curves, and Expansion Velocity of Type I Supernovae as Functions of the Rate of Brightness Decline," *Soviet Astronomy* 21 (1977): 675–682.
33 J. H. Elias et al., "Type I Supernovae in the Infrared and Their Use as Distance Indicators," *Astrophysical Journal* 296 (1985): 379–389.
34 M. M. Phillips, "Type Ia Supernovae as Distance Indicators," in *Supernovae as Cosmological Lighthouses* (Padua, Italy: ASP, 2004).
35 A. G. Riess et al., "Observational Evidence from Supernovae for an Accelerating Universe and a Cosmological Constant," *Astronomical Journal* 116 (1998): 1009–1038.
36 S. Perlmutter et al., "Measurements of Omega and Lambda from 42 High-Redshift Supernovae," *Astrophysical Journal* 517 (1999): 565–586.

Chapter 3

37 A. Sandage, personal interviews with the author, 2007–2010.
38 A. R. Sandage and M. Schwarzschild, "Inhomogeneous Stellar Models. II. Models with Exhausted Cores in Gravitational Contraction," *Astrophysical Journal* 116 (1952): 463.
39 J. A. Frogel, J. G. Cohen, and S. E. Persson, "Globular Cluster Giant Branches and the Metallicity Scale," *Astrophysical Journal* 275 (1983): 773–789.
40 P. W. Merrill, "Technetium in the N-Type Star 19 PISCIUM," *Publications of the Astronomical Society of the Pacific* 68 (1956): 70.
41 A. J. Deutsch, "The Circumstellar Envelope of Alpha Herculis," *Astrophysical Journal* 123 (1956): 210.
42 E. E. Salpeter, "Statistics of Stellar Evolution," *Ricerche Astronomiche* 5 (1958): 231.
43 D. Reimers, "Circumstellar Absorption Lines and Mass Loss from Red Giants," *Mémoires de la Société Royale des Sciences de Liège* 8 (1975): 369–382.
44 W. Baade and F. Zwicky, "Photographic Light-Curves of the Two Supernovae in IC 4182 and NGC 1003," *Astrophysical Journal* 88 (1938): 411.

45 F. Hoyle, "On Nuclear Reactions Occuring in Very Hot Stars. I. The Synthesis of Elements from Carbon to Nickel," *Astrophysical Journal Supplement Series* 1 (1954): 121.

46 W. A. Fowler, "Nuclear Astrophysics—Today and Yesterday," *Engineering and Science* (Caltech) (1969).

47 W. Baade et al., "Supernovae and Californium 254," *Publications of the Astronomical Society of the Pacific* 68 (1956): 296.

48 E. M. Burbidge et al., "Synthesis of the Elements in Stars," *Reviews of Modern Physics* 29 (1957): 547–650.

49 D. A. Coulter et al., "Swope Supernova Survey 2017a (SSS17a), the Optical Counterpart to a Gravitational Wave Source," *Science* 358 (2017): 1556–1558; A. Piro, private communication, 2019.

50 H. L. Helfer, G. Wallerstein, and J. L. Greenstein, "Abundances in Some Population. II. K Giants." *Astrophysical Journal* 129 (1959): 700.

51 J. L. Greenstein and V. L. Trimble, "The Einstein Redshift in White Dwarfs," *Astrophysical Journal* 149 (1967): 283.

52 J. L. Greenstein, J. B. Oke, and H. L. Shipman, "Effective Temperature, Radius, and Gravitational Redshift of Sirius B," *Astrophysical Journal* 169 (1971): 563.

Chapter 4

53 J. W. Chamberlain and L. H. Aller, "The Atmospheres of A-Type Subdwarfs and 95 Leonis," *Astrophysical Journal* 114 (1951): 52.

54 N. G. Roman, "A Group of High Velocity F-Type Stars," *Astronomical Journal* 59 (1954): 307–312.

55 A. Sandage, personal interviews with the author, 2007–2010.

56 A. R. Sandage and M. F. Walker, "The Globular Cluster NGC 4147," *Astronomical Journal* 60 (1955): 230.

57 O. J. Eggen, "Space Motions and Distribution of the Apparently Bright B-Type Stars," *Royal Observatory Bulletin* 41 (1961): 245–287; O. J. Eggen, "Space-Velocity Vectors for 3483 Stars with Accurately Determined Proper Motion and Radial Velocity," *Royal Observatory Bulletin* 51 (1962).

58 A. Sandage, personal interviews with the author, 2007–2010.

59 A. R. Sandage and O. J. Eggen, "On the Existence of Subdwarfs in the (M Bol, log Te)-Diagram," *Monthly Notices of the Royal Astronomical Society* 119 (1959): 278.

60   D. Lynden-Bell, personal interviews with the author, 2008–2011.
61   Sandage and Walker, "The Globular Cluster NGC 4147."
62   H. L. Johnson and A. R. Sandage, "Three-Color Photometry in the Globular Cluster M3," *Astrophysical Journal* 124 (1956): 379.
63   E. M. Burbidge et al., "Synthesis of the Elements in Stars," *Reviews of Modern Physics* 29 (1957): 547–650.
64   A. Sandage, personal interviews with the author, 2007–2010.
65   L. Searle and R. Zinn, "Compositions of Halo Clusters and the Formation of the Galactic Halo," *Astrophysical Journal* 225 (1978): 357–379.
66   B. V. Kukarkin, *Gobular Star Clusters. The General Catalogue of Globular Star Clusters of Our Galaxy, Concerning Information on 129 Objects Known before 1974* (Moscow: Sternberg State Astron. Inst., 1974).
67   Kukarkin, *Gobular Star Clusters*.
68   L. Searle and R. Zinn, "Compositions of Halo Clusters and the Formation of the Galactic Halo," *Astrophysical Journal* 225 (1978): 357–379.
69   D. Geisler et al., "Chemical Abundances and Kinematics in Globular Clusters and Local Group Dwarf Galaxies and Their Implications for Formation Theories of the Galactic Halo," *Publications of the Astronomical Society of the Pacific* 119 (2007): 939–961.
70   O. Eggen, "Moving Groups of Stars," in *Galactic Structure*, ed. A. Blaauw and M. Schmidt (Chicago: University of Chicago Press, 1965), 111–129.
71   Searle and Zinn, "Compositions of Halo Clusters and the Formation of the Galactic Halo."

Chapter 5

72   F. Zwicky and M. A. Zwicky, *Catalogue of Selected Compact Galaxies and of Post-eruptive Galaxies* (Gümligen: F. Zwicky, 1971).
73   F. Zwicky, "Multiple Galaxies," *Ergebnisse der exakten Naturwissenschaften* 29 (1956): 344–385.
74   F. Zwicky, "Luminous and Dark Formations of Intergalactic Matter," *Physics Today* 6 (1953): 7; F. Zwicky, "Contributions to Applied Mechanics and Related Subjects," in *Theodore von Kármán Anniversary* (Pasadena: California Institute of Technology, 1941).
75   Zwicky, "Multiple Galaxies" (1956).
76   Zwicky, "Contributions to Applied Mechanics and Related Subjects."

77   Zwicky, "Multiple Galaxies" (1956); F. Zwicky, "Multiple Galaxies," *Handbuch der Physik* 53 (1959): 373.
78   W. Baade and R. Minkowski, "On the Identification of Radio Sources," *Astrophysical Journal* 119 (1954): 215.
79   L. Spitzer Jr. and W. Baade, "Stellar Populations and Collisions of Galaxies," *Astrophysical Journal* 113 (1951): 413.
80   Zwicky, "Multiple Galaxies" (1959).
81   A. Sandage, "Photoelectric Observations of the Interacting Galaxies VV 117 and VV 123 Related to the Time of Formation of Their Satellites," *Astrophysical Journal* 138 (1963): 863.
82   B. A. Vorontsov-Velyaminov, *Atlas and Catalogue of Interacting Galaxies* (Moscow: Sternberg Astronomical Institute, Moscow State University, 1959).
83   H. Arp, personal interview with the author, 2009.
84   A. Sandage, *The Hubble Atlas of Galaxies* (Washington: Carnegie Institution, 1961).
85   E. Hubble, *The Realm of the Nebulae* (New Haven: Yale University Press, 1936).
86   H. Arp, personal interview with the author, 2009.
87   H. Arp, *Atlas of Peculiar Galaxies* (Pasadena: California Institute of Technology, 1966); also published as H. Arp, "Atlas of Peculiar Galaxies," *Astrophysical Journal Supplement Series* 14 (1966): 1.
88   H. Arp, personal interview with the author, 2009.
89   A. Toomre, personal interviews with the author, 2009–2014.
90   A. Toomre, personal interviews with the author, 2009–2014.
91   A. Toomre, "Spiral Waves Caused by a Passage of the Lmc?," in *The Spiral Structure of Our Galaxy*, ed. W. Becker and G. Contopoulos (Dordrecht: Reidel, 1970).
92   A. Toomre and J. Toomre, "Galactic Bridges and Tails," *Astrophysical Journal* 178 (1972): 623–666.
93   A. Toomre and J. Toomre, "Model of the Encounter between NGC 5194 and 5195," *Bulletin of the American Astronomical Society* (1972).
94   Toomre and Toomre, "Model of the Encounter between NGC 5194 and 5195."
95   J. C. Theys and E. A. Spiegel, "Ring Galaxies. I," *Astrophysical Journal* 208 (1976): 650–661.
96   R. Lynds and A. Toomre, "On the Interpretation of Ring Galaxies: The Binary Ring System II Hz 4.," *Astrophysical Journal* 209 (1976): 382–388.
97   Zwicky, "Contributions to Applied Mechanics and Related Subjects."
98   P. Hickson, "Systematic Properties of Compact Groups of Galaxies," *Astrophysical Journal* 255 (1982): 382–391.

99 J. E. Barnes, "Evolution of Compact Groups and the Formation of Elliptical Galaxies," *Nature* 338 (1989): 123–126.
100 V. C. Rubin, personal interviews with the author, 2008–2010.

Chapter 6

101 A. S. Bennett, "The Revised 3C Catalogue of Radio Sources," *Memoirs of the Royal Astronomical Society* 68 (1962): 163.
102 F. G. Smith, "An Attempt to Measure the Annual Parallax or Proper Motion of Four Radio Stars," *Nature* 168 (1951): 962–963.
103 W. Baade and R. Minkowski, "Identification of the Radio Sources in Cassiopeia, Cygnus A, and Puppis A," *Astrophysical Journal* 119 (1954): 206; W. Baade and R. Minkowski, "On the Identification of Radio Sources," *Astrophysical Journal* 119 (1954): 215.
104 Baade and Minkowski, "On the Identification of Radio Sources," 228.
105 T. A. Matthews, personal interview with the author, 2013.
106 A. Sandage, personal interviews with the author, 2007–2010.
107 M. Schmidt, "Spectrum of a Stellar Object Identified with the Radio Source 3C 286," *Astrophysical Journal* 136 (1962): 684.
108 C. Hazard, M. B. Mackey, and A. J. Shimmins, "Investigation of the Radio Source 3C273 by the Method of Lunar Occultations," *Nature* 197 (1963): 1037.
109 T. Wolfe, *The Right Stuff* (New York: Farrar, Straus and Giroux, 2008), 21.
110 J. B. Oke, "Absolute Energy Distribution in the Optical Spectrum of 3C273," *Nature* 197 (1963): 1040.
111 M. Schmidt, "3C 273: A Star-like Object with Large Red-Shift," *Nature* 197 (1963): 1040.
112 J. L. Greenstein and T. A. Matthews, "Red-Shift of the Unusual Radio Source 3C48," *Nature* 197 (1963): 1041.
113 D. Lynden-Bell and F. Schweizer, "Allan R. Sandage, 18 June 1926–13 November 2010," *Biographical Memoirs of Fellows of the Royal Society* 58 (2012): 245–264; D. Lynden-Bell, personal interviews with the author, 2008–2011.
114 M. Schmidt, personal interviews with the author, 2007–2010.
115 M. Schmidt, personal interviews with the author, 2007–2010.
116 T. A. Matthews and A. R. Sandage, "Optical Identification of 3C 48, 3C 196, and 3C 286 with Stellar Objects," *Astrophysical Journal* 138 (1963): 30.
117 H. J. Smith and D. Hoffleit, "Light Variability and Nature of 3C273," *Astronomical Journal* 68 (1963): 292.

118 M. Schmidt, "Large Redshifts of Five Quasi-Stellar Sources," *Astrophysical Journal* 141 (1965): 1295.
119 M. Schmidt, personal interviews with the author, 2007–2010.
120 A. Sandage, "The Existence of a Major New Constituent of the Universe: The Quasistellar Galaxies," *Astrophysical Journal* 141 (1965): 1560.
121 M. Schmidt, personal interviews with the author, 2007–2010.
122 J. E. Gunn, "On the Distances of the Quasi-Stellar Objects," *Astrophysical Journal* 164 (1971): L113.
123 A. Sandage, "The Redshift-Distance Relation. I. Angular Diameter of First Ranked Cluster Galaxies as a Function of Redshift: The Aperture Correction to Magnitudes," *Astrophysical Journal* 173 (1972): 485.
124 J. Kristian, "Quasars as Events in the Nuclei of Galaxies: The Evidence from Direct Photographs," *Astrophysical Journal* 179 (1973): L61.
125 J. B. Oke and J. E. Gunn, "An Efficient Low Resolution and Moderate Resolution Spectrograph for the Hale Telescope," *Publications of the Astronomical Society of the Pacific* 94 (1982): 586.
126 T. A. Boroson and J. B. Oke, "Detection of the Underlying Galaxy in the QSO 3C48," *Nature* 296 (1982): 397–399.
127 M. Schmidt, personal interviews with the author, 2007–2010.
128 M. Schmidt, "Space Distribution and Luminosity Functions of Quasi-Stellar Radio Sources," *Astrophysical Journal* 151 (1968) 393.
129 R. F. Green, M. Schmidt, and J. Liebert, "The Palomar-Green Catalog of Ultraviolet-Excess Stellar Objects," *Astrophysical Journal Supplement Series* 61 (1986): 305–352.
130 M. Schmidt and R. F. Green, "Quasar Evolution Derived from the Palomar Bright Quasar Survey and Other Complete Quasar Surveys," *Astrophysical Journal* 269 (1983): 352–374.
131 M. Schmidt, personal interviews with the author, 2007–2010.
132 M. Schmidt, D. P. Schneider, and J. E. Gunn, "Spectroscopic CCD Surveys for Quasars at Large Redshift. IV. Evolution of the Luminosity Function from Quasars Detected by Their Lyman-Alpha Emission," *Astronomical Journal* 110 (1995): 68.

Chapter 7

133 G. Neugebauer, D. E. Martz, and R. B. Leighton, "Observations of Extremely Cool Stars," *Astrophysical Journal* 142 (1965): 399–401.

134 G. Neugebauer and R. B. Leighton, *Two-Micron Sky Survey—A Preliminary Catalog* (Washington, DC: NASA SP-3047, Government Printing Office, 1969), 309.

135 E. Becklin, personal interviews with the author, 2008–2019.

136 E. E. Becklin and G. Neugebauer, "Observations of an Infrared Star in the Orion Nebula," *Astrophysical Journal* 147 (1967): 799.

137 E. E. Becklin et al., "The Unusual Infrared Object IRC+10216," *Astrophysical Journal* 158 (1969): L133.

138 R. I. Toombs et al., "Infrared Diameter of IRC+10216 Determined from Lunar Occultations," *Astrophysical Journal* 173 (1972): L71.

139 E. E. Becklin and G. Neugebauer, "Infrared Observations of the Galactic Center," *Astrophysical Journal* 151 (1968): 145.

140 E. Becklin, personal interviews with the author, 2008–2019.

141 A. R. Sandage, E. E. Becklin, and G. Neugebauer, "UBVRIHKL Photometry of the Central Region of M31," *Astrophysical Journal* 157 (1969): 55.

142 E. Becklin, personal interviews with the author, 2008–2019.

143 E. E. Becklin and G. Neugebauer, "High-Resolution Maps of the Galactic Center at 2.2 and 10 Microns," *Astrophysical Journal* 200 (1975): L71–L74.

144 E. E. Becklin et al., "The Size of NGC 1068 at 10 Microns," *Astrophysical Journal* 186 (1973): L69.

145 B. T. Soifer et al., "The IRAS Bright Galaxy Sample. II—The Sample and Luminosity Function," *Astrophysical Journal* 320 (1987): 238–257.

146 B. T. Soifer et al., "The Luminosity Function and Space Density of the Most Luminous Galaxies in the IRAS Survey," *Astrophysical Journal* 303 (1986): L41–L44.

147 M. Schmidt and R. F. Green, "Quasar Evolution Derived from the Palomar Bright Quasar Survey and Other Complete Quasar Surveys," *Astrophysical Journal* 269 (1983): 352–374.

148 Soifer et al., "The IRAS Bright Galaxy Sample. II."

149 F. Zwicky and M. A. Zwicky, *Catalogue of Selected Compact Galaxies and of Post-eruptive Galaxies* (Gümligen: F. Zwicky, 1971).

150 Soifer et al., "The Luminosity Function and Space Density of the Most Luminous Galaxies in the IRAS Survey."

Chapter 8

151   H. Arp, "Companion Galaxies on the Ends of Spiral Arms," *Astronomy and Astrophysics* 3 (1969): 418–435.
152   F. Zwicky, "Blue Compact Galaxies," *Astrophysical Journal* 142 (1965): 1293.
153   W. L. W. Sargent, "A Spectroscopic Survey of Compact and Peculiar Galaxies," *Astrophysical Journal* 160 (1970): 405.
154   F. Zwicky and M. A. Zwicky, *Catalogue of Selected Compact Galaxies and of Post-eruptive Galaxies* (Gümligen: F. Zwicky, 1971).
155   Sargent, "A Spectroscopic Survey of Compact and Peculiar Galaxies."
156   W. L. W. Sargent and L. Searle, "Isolated Extragalactic H II Regions," *Astrophysical Journal* 162 (1970): L155.
157   L. Searle and W. L. W. Sargent, "Inferences from the Composition of Two Dwarf Blue Galaxies," *Astrophysical Journal* 173 (1972): 25.
158   L. Searle, W. L. W. Sargent, and W. G. Bagnuolo, "The History of Star Formation and the Colors of Late-Type Galaxies," *Astrophysical Journal* 179 (1973): 427–438.
159   Searle, Sargent, and Bagnuolo, "The History of Star Formation and the Colors of Late-Type Galaxies."
160   H. Arp and A. Sandage, "Spectra of the Two Brightest Objects in the Amorphous Galaxy NGC 1569—Superluminous Young Star Clusters—or Stars in a Nearby Peculiar Galaxy?" *Astronomical Journal* 90 (1985): 1163–1171.
161   C. R. Lynds and A. Sandage, "Evidence for an Explosion in the Center of the Galaxy M82," *Astrophysical Journal* 137 (1963): 137.
162   N. Visvanathan and A. Sandage, "Linear Polarization of the Hα Emission Line in the Halo of M82 and the Radiation Mechanism of the Filaments," *Astrophysical Journal* 176 (1972): 57.
163   R. Lynds and A. Toomre, "On the Interpretation of Ring Galaxies: The Binary Ring System II Hz 4," *Astrophysical Journal* 209 (1976): 382–388.
164   C. K. Xu, personal interviews with the author, 2010.
165   C. K. Xu et al., "Physical Conditions and Star Formation Activity in the Intragroup Medium of Stephan's Quintet," *Astrophysical Journal* 595 (2003): 665–684.
166   P. N. Appleton, personal interview with the author, 2010.
167   J. R. Graham et al., "The Double Nucleus of Arp 220 Unveiled," *Astrophysical Journal* 354 (1990): L5–L8.
168   D. Lynden-Bell, "Galactic Nuclei as Collapsed Old Quasars," *Nature* 223 (1969): 690–694.

169   P. J. Young et al., "Evidence for a Supermassive Object in the Nucleus of the Galaxy M87 from SIT and CCD Area Photometry," *Astrophysical Journal* 221 (1978): 721–730.

170   A. Dressler, personal interviews with the author, 2010–2014.

171   A. Dressler, "Studying the Internal Kinematics of Galaxies Using the Calcium Infrared Triplet," *Astrophysical Journal* 286 (1984): 97–105.

172   A. Dressler and D. O. Richstone, "Stellar Dynamics in the Nuclei of M31 and M32—Evidence for Massive Black Holes?," *Astrophysical Journal* 324 (1988): 701–713.

173   A. V. Filippenko and W. L. W. Sargent, "A Search for 'Dwarf' Seyfert 1 Nuclei. I—The Initial Data and Results," *Astrophysical Journal Supplement Series* 57 (1985): 503–522.

174   L. C. Ho, personal interview with the author, 2010.

175   L. Ho, "Supermassive Black Holes in Galactic Nuclei: Observational Evidence and Astrophysical Consequences," in *Observational Evidence for Black Holes in the Universe*, ed. S. K. Chakrabarti (Dordrecht: Springer, 1999).

176   M. Peimbert and S. Torres-Peimbert, "Physical Conditions in the Nucleus of M81," *Astrophysical Journal* 245 (1981): 845–856.

177   A. V. Filippenko and W. L. W. Sargent, "A Detailed Study of the Emission Lines in the Seyfert 1 Nucleus of M81," *Astrophysical Journal* 324 (1988): 134–153.

178   L. C. Ho, A. V. Filippenko, and W. L. W. Sargent, "New Insights into the Physical Nature of LINERs from a Multiwavelength Analysis of the Nucleus of M81," *Astrophysical Journal* 462 (1996): 183.

Chapter 9

179   M. Schmidt, "Large Redshifts of Five Quasi-Stellar Sources," *Astrophysical Journal* 141 (1965): 1295.

180   J. E. Gunn and B. A. Peterson, "On the Density of Neutral Hydrogen in Intergalactic Space," *Astrophysical Journal* 142 (1965): 1633–1641.

181   J. N. Bahcall and E. E. Salpeter, "On the Interaction of Radiation from Distant Sources with the Intervening Medium," *Astrophysical Journal* 142 (1965): 1677–1680.

182   J. L. Greenstein and M. Schmidt, "The Two Absorption-Line Redshifts in Parkes 0237-23," *Astrophysical Journal* 148 (1967): L13.

183   R. Lynds, "The Absorption-Line Spectrum of 4c 05.34," *Astrophysical Journal* 164 (1971): L73.

184 A. Boksenberg, personal interviews with the author, 2011–2013.
185 W. L. W. Sargent, personal interviews with the author, 2008–2011.
186 A. Boksenberg, personal interviews with the author, 2011–2013.
187 W. L. W. Sargent et al., "The Distribution of Lyman-Alpha Absorption Lines in the Spectra of Six QSOs—Evidence for an Intergalactic Origin," *Astrophysical Journal Supplement Series* 42 (1980): 41–81.
188 M. Rauch, private communication, 2019.
189 D. A. Frail et al., "The Radio Afterglow from the γ-ray Burst of 8 May 1997," *Nature* 389 (1997): 261–263.
190 S. G. Djorgovski et al., "The Optical Counterpart to the γ-ray Burst GRB970508," *Nature* 387 (1997): 876–878; M. R. Metzger et al., "Spectral Constraints on the Redshift of the Optical Counterpart to the γ-ray Burst of 8 May 1997," *Nature* 387 (1997): 878–880.

Chapter 10

191 E. P. Hubble, *Realm of the Nebulae* (New Haven: Yale University Press, 1936).
192 J. E. Gunn and J. A. Westphal, "Care Feeding and Use of Charge-Coupled Device / CCD / Imagers at Palomar Observatory," *Society of Photo-Optical Instrumentation Engineers (SPIE) Conference Series* (1981): 16.
193 A. Dressler, J. E. Gunn, and D. P. Schneider, "Spectroscopy of Galaxies in Distant Clusters. III—The Population of CL 0024 + 1654," *Astrophysical Journal* 294 (1985): 70–80; A. Dressler, personal interviews with the author, 2010–2014.
194 L. Spitzer Jr. and W. Baade, "Stellar Populations and Collisions of Galaxies," *Astrophysical Journal* 113 (1951): 413.
195 J. E. Gunn and J. R. Gott III, "On the Infall of Matter into Clusters of Galaxies and Some Effects on Their Evolution," *Astrophysical Journal* 176 (1972): 1.
196 J. E. Gunn, personal interviews with the author, 2008–2017.
197 C. C. Steidel and W. L. W. Sargent, "Mg II Absorption in the Spectra of 103 QSOs—Implications for the Evolution of Gas in High-Redshift Galaxies," *Astrophysical Journal Supplement Series* 80 (1992): 1–108; W. L. W. Sargent, A. Boksenberg, and C. C. Steidel, "C IV Absorption in a New Sample of 55 QSOs—Evolution and Clustering of the Heavy-Element Absorption Redshifts," *Astrophysical Journal Supplement Series* 68 (1988): 539–641.
198 J. Bergeron and P. Boissé, "A Sample of Galaxies Giving Rise to Mg II Quasar Absorption Systems," *Astronomy and Astrophysics* 243 (1991): 344–366.

199 W. A. Baum, "Photoelectric Determinations of Redshifts Beyond 0.2 c," *Astronomical Journal* 62 (1957): 6–7.

200 W. A. Baum, "Photoelectric Magnitudes and Red-Shifts," *International Astronomical Union Symposium* 15 (1962): 390–400.

201 C. C. Steidel et al., "Spectroscopic Confirmation of a Population of Normal Star-Forming Galaxies at Redshifts Z > 3," *Astrophysical Journal* 462 (1996): L17.

202 C. C. Steidel et al., "A Large Structure of Galaxies at Redshift Z ~ 3 and Its Cosmological Implications," *Astrophysical Journal* 492 (1998): 428; K. L. Adelberger et al., "A Counts-in-Cells Analysis of Lyman-Break Galaxies at Redshift Z ~ 3," *Astrophysical Journal* 505 (1998): 18–24.

203 C. C. Steidel, personal interviews with the author, 2009–2011.

204 A. E. Shapley, personal interview with the author, 2011.

205 Steidel et al., "A Large Structure of Galaxies at Redshift Z ~ 3 and Its Cosmological Implications."

206 C. C. Steidel et al., "Lyman-α Imaging of a Proto-cluster Region at <z> = 3.09," *Astrophysical Journal* 532 (2000): 170–182.

Chapter 11

207 B. C. Murray, personal interviews with the author, 2008–2010.

208 B. C. Murray, personal interviews with the author, 2008–2010.

209 B. C. Murray, personal interviews with the author, 2008–2010.

210 R. L. Wildey, B. C. Murray, and J. A. Westphal, "Thermal Infrared Emission of the Jovian Disk," *Journal of Geophysical Research* 70 (1965): 3711.

211 F. C. Gillett and J. A. Westphal, "Observations of 7.9-Micron Limb Brightening on Jupiter," *Astrophysical Journal* 179 (1973): L153.

212 F. C. Gillett, F. J. Low, and W. A. Stein, "The 2.8–14-Micron Spectrum of Jupiter," *Astrophysical Journal* 157 (1969): 925.

213 J. A. Westphal, "Observations of Localised 5-Micron Radiation from Jupiter," *Astrophysical Journal* 157 (1969): L63–L64.

214 J. A. Westphal, K. Matthews, and R. J. Terrile, "Five-Micron Pictures of Jupiter," *Astrophysical Journal* 188 (1974): L111–L112; R. J. Terrile, "High Spatial Resolution Infrared Imaging of Jupiter: Implications for the Vertical Cloud Structure from Five-Micron Measurements," PhD thesis, California Institue of Technology, 1978, 1.

215 R. J. Terrile and J. A. Westphal, "The Vertical Cloud Structure of Jupiter from 5 μm Measurements," *Icarus* 30 (1977): 274–281.

216 R. J. Terrile, personal interview with the author, 2008.
217 R. J. Terrile et al., "Infrared Images of Jupiter at 5-Micrometer Wavelength During the Voyager 1 Encounter," *Science* 204 (1979): 948.
218 R. J. Terrile, personal interview with the author, 2008.
219 C. S. Shoemaker, "Twelve Years on the Palomar 18-Inch Schmidt," *Journal of the Royal Astronomical Society of Canada* 90 (1996): 18.
220 C. Shoemaker, personal interview with the author, 2008.
221 D. Banfield et al., "2 μm Spectrophotometry of Jovian Stratospheric Aerosols—Scattering Opacities, Vertical Distributions, and Wind Speeds," *Icarus* 121 (1996): 389–410.
222 P. D. Nicholson et al., "Palomar Observations of the R Impact of Comet Shoemaker-Levy 9: II. Spectra," *Geophysical Research Letters* 22 (1995): 1617–1620.
223 R. S. Richardson, "A New Asteroid with Smallest Known Mean Distance," *Publications of the Astronomical Society of the Pacific* 61 (1949): 162.
224 D. Jewitt and J. Luu, "Discovery of the Candidate Kuiper Belt Object 1992 QB1," *Nature* 362 (1993): 730–732.
225 M. Brown, personal interviews with the author, 2008–2019.
226 E. F. Helin, S. H. Pravdo, K. H. Lawrence, and M. D. Hicks, "The Near-Earth Asteroid Tracking (NEAT) Program," *Bulletin of the American Astronomical Society* 32 (2000): 750.
227 C. A. Trujillo and M. E. Brown, "The Caltech Survey for the Brightest Kuiper Belt Objects," *Bulletin of the American Astronomical Society* 35 (2003): 1015.
228 M. E. Brown, C. Trujillo, and D. Rabinowitz, "Discovery of a Candidate Inner Oort Cloud Planetoid," *Astrophysical Journal* 617 (2004): 645–649.
229 M. E. Brown, C. A. Trujillo, and D. L. Rabinowitz, "Discovery of a Planetary-Sized Object in the Scattered Kuiper Belt," *Astrophysical Journal* 635 (2005): L97–L100.
230 M. E. Brown et al., "A Collisional Family of Icy Objects in the Kuiper Belt," *Nature* 446 (2007): 294–296.
231 M. Brown, personal interviews with the author, 2008–2019.
232 Brown et al., "A Collisional Family of Icy Objects in the Kuiper Belt."
233 Brown, Trujillo, and Rabinowitz, "Discovery of a Planetary-Sized Object in the Scattered Kuiper Belt."
234 K. M. Barkume, M. E. Brown, and E. L. Schaller, "Near Infrared Spectroscopy of Icy Planetoids," *Bulletin of the American Astronomical Society* 37 (2005): 738.

235   C. T. Kowal, "Chiron," in *Asteroids*, ed. T. Gehrels with M. S. Matthews (Tucson: University of Arizona Press, 1979), 436–439.
236   K. Batygin, private communication.
237   M. Brown and K. Batygin, "Observational Constraints on the Orbit and Location of Planet Nine in the Outer Solar System," *Astrophysical Journal Letters* 824 (2016): 2.

CHAPTER 12

238   F. Zwicky, "Die Rotverschiebung von extragalaktischen Nebeln," *Helvetica Physica Acta* 6 (1933): 110–127.
239   S. S. Kumar, "Models for Stars of Very Low Mass," Institute for Space Studies Report X-644-62-78 (1962), 1; S. S. Kumar, "The Structure of Stars of Very Low Mass," *Astrophysical Journal* 137 (1963): 1121.
240   J. C. Tarter, personal interview with the author, 2015.
241   J. C. Tarter, "The Interaction of Gas and Galaxies within Galaxy Clusters," PhD thesis, University of California, Berkeley, 1975, 1.
242   M. Clampin et al., "High Speed Quadrant CCDs for Adaptive Optics," in *CCDs in Astronomy: Proceedings of a Conference Held in Tucson, Arizona, 6–8 September 1989*, ed. G. H. Jacoby (San Francisco: Astronomical Society of the Pacific, 1990), 367–373.
243   T. Nakajima et al., "A Coronagraphic Search for Brown Dwarfs around Nearby Stars," *Astrophysical Journal* 428 (1994): 797–804.
244   K. Matthews, personal interviews with the author, 2008–2010.
245   T. Tsuji, "Molecular Abundance in Stellar Atmospheres," *Annals of the Tokyo Astronomical Observatory* 9 (1964).
246   T. Tsuji, R. Blomme, and N. Grevesse, "Molecules in the Atmospheres of Brown Dwarfs," *Laboratory and Astronomical High Resolution Spectra* 81 (1995): 566.
247   Tsuji, Blomme, and Grevesse, "Molecules in the Atmospheres of Brown Dwarfs."
248   T. Nakajima et al., "Discovery of a Cool Brown Dwarf," *Nature* 378 (1995): 463–465.
249   B. R. Oppenheimer et al., "Infrared Spectrum of the Cool Brown Dwarf Gl 229B," *Science* 270 (1995): 1478–1479.
250   A. J. Burgasser, "The Discovery and Characterization of Methane-Bearing Brown Dwarfs and the Definition of the T Spectral Class," PhD thesis, California Institute of Technology, 2001, 116; A. J. Burgasser, personal interviews with the author, 2015–2019.

251 Burgasser, "The Discovery and Characterization of Methane-Bearing Brown Dwarfs and the Definition of the T Spectral Class."

252 C. Marois, B. Zuckerman, Q. M. Konopacky, B. Macintosh, and T. Barman, "Images of a Fourth Planet Orbiting HR 8799," *Nature* 468 (2010): 1080–1083.

253 B. R. Oppenheimer et al., "Reconnaissance of the HR 8799 Exosolar System. I. Near-Infrared Spectroscopy," *Astrophysical Journal* 768 (2013): 24.

254 A. Rau et al., "Exploring the Optical Transient Sky with the Palomar Transient Factory," *Publications of the Astronomical Society of the Pacific* 121 (2009): 1334–1351.

255 N. M. Law et al., "The Palomar Transient Factory: System Overview, Performance, and First Results," *Publications of the Astronomical Society of the Pacific* 121 (2009): 1395–1408.

256 A. L. Piro, P. Chang, and N. N. Weinberg, "Shock Breakout from Type Ia Supernova," *Astrophysical Journal* 708 (2010): 598; A. L. Piro, private communication, 2019.

257 W. Li et al., "Exclusion of a Luminous Red Giant as a Companion Star to the Progenitor of Supernova SN 2011fe," *Nature* 480 (2011): 348–350; P. E. Nugent et al., "Supernova SN 2011fe from an Exploding Carbon-Oxygen White Dwarf Star," *Nature* 480 (2011): 344–347.

258 M. L. Graham et al., "Constraining the Progenitor Companion of the Nearby Type Ia SN 2011fe with a Nebular Spectrum at +981 d," *Monthly Notices of the Royal Astronomical Society* 454 (2015): 1948–1957.

259 R. M. Quimby et al., "Hydrogen-Poor Superluminous Stellar Explosions," *Nature* 474 (2011): 487–489.

260 E. C. Bellm et al., "The Zwicky Transient Facility: System Overview, Performance, and First Results," *Publications of the Astronomical Society of the Pacific* 131 (2019): 018002; M. J. Graham et al., "The Zwicky Transient Facility: Science Objectives," *Publications of the Astronomical Society of the Pacific* 131 (2019): 078001.

261 C. Payne-Gaposchkin, "A Special Kind of Astronomer," *Sky and Telescope* 47 (1974): 311.

262 K. B. Burdge, "General Relativistic Orbital Decay in a Seven-Minute-Orbital-Period Eclipsing Binary System," *Nature* 571 (2019): 528–531.

INDEX

Abell, George, 32
active galactic nuclei (AGNs), 133, 179–180, 210–212
adaptive optics, 198, 243–245, 247, 257
adaptive optics coronagraph (AOC), 244–247, 249, 256
Adelberger, Kurt, 207, 209, 211
Air Force, US, 63, 122, 137, 215
Aller, Lawrence, 71
Andromeda galaxy. *See* Messier 31
Antennae galaxies (NGC 4038/4039), 102–103
Appleton, Philip, 171
Arp, Halton (Chip), xiii, 27, 95–100, 106, 110, 145, 163–165, 185, 197. *See also* Atlas of Peculiar Galaxies
Arp 220, galaxy, 154, 171–173
Atlas and Catalogue of Interacting Galaxies (Vorontsov-Velyaminov), 95
Atlas of Peculiar Galaxies (Arp), 97–100, 101, 102, 104, 153, 154, 159, 163, 172

B²FH (Burbidge, Burbidge, Fowler, and Hoyle paper on synthesis of elements in stars), 61–63, 75, 82
Baade, Walter, 11–12, 22–27, 34, 38, 48–51, 82, 95, 118, 121, 148, 161, 164–165, 202, 227. *See also* stellar populations I and II
    distance to M31, 23–27
    on galactic collisions, 92–94
    identification of radio sources, 113–118
    stellar evolution, 48–51
    supernovae, 36, 58, 61, 260
Bagnuolo, William, 161
Bahcall, John N., xi, 185, 189
Balmer lines of hydrogen, 36, 64, 125, 159, 201, 266
Barkume, Kristina, 233, 234
Barnes, Joshua, 106–108
baryons, 182, 190, 193, 196, 212
Basri, Gibor, 243
Batygin, Konstantin, 236–238
Baum, William A., 205–206
Bautz-Morgan effect, 32–34
Becklin, Eric, 143–145, 147–149
Becklin-Neugebauer Object, 144–145, 149
BeppoSAX, satellite, 191–192
Bergeron, Jacqueline, 205
big bang, 135, 160, 193, 196, 209
    elements synthesized in, 55–56, 59, 64, 160, 211
    theory, 182
Big Eye. *See* Palomar Observatory: 200-inch telescope
Bikini Atoll, 60–61
black holes, stellar-mass, xii, 67, 112, 153, 196, 242, 265
black holes, supermassive (SMBH), xi, 17, 93, 112, 117, 138, 149, 158, 159, 190, 260

correlation of mass with bulge mass of galaxy, 180
observational evidence for, 173–174, 175–179
blobs. *See* Lyman-α blobs
Bloom, Joshua, 262
blue compact dwarf galaxies, 160–162
Boksenberg, Alexander, 174, 186–189
Boksenberg's Flying Circus, 187–189
bolometric luminosity, 151–152, 153, 172
Bolton, John, 118, 119
Boroson, Todd, 132–133
Bowen, Ira, 25, 56, 59, 97, 121, 183
Brown, Michael, 228–230, 233–238
brown dwarf stars, 243, 245, 247–252, 254, 256, 258, 268. *See also* Gliese 229B
Burbidge, E. Margaret, 59, 61, 62, 75, 94, 185. *See also* B²FH
Burbidge, Geoffrey, 59, 61, 75, 94. *See also* B²FH
Burdge, Kevin, 267
Burgasser, Adam, 250–251
Butcher, Harvey, 197–198, 200, 202

Californium-254, 60–61
carbon star, 145–146
Carrasco, Juan, 199
Cartwheel galaxy, 104–105
discovery by Zwicky, 91
Cas A, supernova remnant, 240
Centaurs (comet-like asteroids), 236
Centaurus A, radio source, 93. *See also* NGC 5128

Cepheid variable stars, 21–27, 30–31, 38
light curves of, 23, 24, 26
period-luminosity relation for types I and II, 27
Cerro Tololo Inter-American Observatory, 42, 43, 207
Chamberlain, Joseph, 71
chemical elements, 48, 59. *See also* helium
abundances of, in stars, 64, 71, 78–80, 160, 189
in brown dwarf stars, 248, 251
creation during fusion process in stars, 52, 56, 59–60, 64, 75, 173
creation in neutron stars, 61–62
creation in supernova explosions, 60–62, 77, 265
dispersal of, 55–56, 76–77, 82, 146, 165, 211–212, 265
Chiron, 236. *See also* Centaurs
Chiu, Hong-Yee, 129
Christy, Robert, 61
Cohen, Judith, 55
collisions, between galaxies, 92–94, 99, 109, 116, 137, 158–159, 164, 169–172, 202
modeling of, 100–106
collisions, within the solar system, 223–227, 232–234, 236
color-magnitude diagram for stars, 32, 48–55, 63, 74, 79–80
composite, 52–55
of Messier 3, 48–52
cosmic web, 182, 183, 193, 206, 212
Cosmological Principle, 201
crystal stars, 67

Cygnus A, radio source, 113–119
  optical counterpart, 114–115
  radio lobes of, 117

Dark energy, 6, 18, 196, 212, 265
dark matter, 18, 106–110, 242, 252, 258, 268, 269
  in cosmological models, 193, 196, 211, 212, 241–242
  halos, 70, 82, 84, 107, 108
  in simulations, 106–108
Deutsch, Armin J., 57–58, 60
Djorgovski, S. George, 191–192
double spectrograph, 132, 165, 170, 175, 177, 199
Dressler, Alan, 175–176, 198, 200–203
DuBridge, Lee A., 8
Dunlop, James, 92
Durrance, Samuel, 243–245, 256
dwarf galaxies, 58, 160–162, 201, 204, 261, 266
  and formation of galaxy halos, 81–84
dwarf stars. *See* brown dwarf stars; subdwarf stars; white dwarf stars

Eggen, Olin, 65, 73–77, 197. *See also* ELS; "moving groups" of stars
Einstein, Albert, 6, 9, 21, 66, 269
Einstein ring, 269
electromagnetic spectrum, xii, 17, 62, 140, 151, 184, 190, 191
electron degeneracy, 65, 242
Elias, Jonathan H. (Jay), 41–42
ELS (Eggen, Lynden-Bell, and Sandage paper on formation of Milky Way), 73–80, 82. *See also* Searle-Zinn model of galaxy formation
Eris, dwarf planet, 234–235, 238
ESO 137-001, galaxy, 203
exoplanets, 238, 253–258, 259, 261, 268
exploding stars, 11, 58

Ferocci, Marco, 191
Filippenko, Alexei, 176–179, 264
"flashing" galaxies, 161, 163
fly spanker, 49
forbidden absorption bands, 36, 40, 116, 119
Fordham, John, 188
Fosdick, Raymond B., 3
Four-Shooter camera (4-Shooter), 136–137, 198–199
Fowler, William, 59, 61, 75. *See also* B^{2}FH
Frail, Dale, 191
Frogel, Jay, 55

Gaia satellite (European Space Agency), 67
galaxy "birth canal," 94
Galileo, spacecraft and atmospheric entry probe, 221–222
gamma-ray bursts, 191–192, 265
giant stars, 23, 51–58, 63–64, 201
  red giants, 47, 48, 50, 51, 52, 57–61, 65, 67, 77–79, 145, 264–265
  supergiants, 56–57, 58–60, 164
Gliese catalog, 245–246

Gliese 229B, brown dwarf star, 252, 254, 258
  discovery of, 247–250
globular clusters, 23, 27, 32, 48, 63, 64, 72, 76, 78, 79, 82–84. *See also* Messier 3
  estimating ages of, 48, 52, 75, 80
  in halos of galaxies, 80, 81
  ultraviolet excess in, 75
  variable stars in, 48
Goldreich, Peter, 99
Golimowski, David, 243, 245
Gott, Richard, 202–203
Graham, James, 171
gravitational interactions, between galaxies, 81–94, 98, 158–159, 163, 166, 169
  numerical modeling of, 99–106, 110, 197
gravitational redshifts, 66, 126
gravitational waves, x, 61, 62, 267
Green, Richard, 134–136, 206. *See also* Palomar-Green catalog
Greenstein, Jesse, 62, 72, 73, 118, 126–127, 185, 187
  chemical abundances, 63–64
  white dwarf stars, 65–67
grism, 136–137
Gunn, James E., 34–35, 131, 132, 184–185, 205, 206, 271
  on galaxy evolution in distant clusters, 198–203
  as instrument designer, 35, 132, 136–137, 198, 200–201, 244
  on intergalactic density of neutral hydrogen, 184–185
  introduction of CCDs to Palomar, 198
  quasar surveys, 136–137, 205, 206
Gunn-Peterson trough, 184–185

Hale, George Ellery, ix, xi, 3–9, 113
  as builder of world's largest telescopes, ix, 5–9
  leader in transition from astronomy to astrophysics, 4
Hale telescope. *See* Palomar Observatory: 200-inch (Hale) telescope
Hamilton, Don, 206
Hardy, Eduardo, 32
Haumea, dwarf planet, 233–234, 235
Hazard, Cyril, 122
Helfer, Lawrence, 63–64
Helin, Eleanor (Glo), 229
helium, 51–52, 65, 189, 265
  origin of, 48, 50, 51, 59, 160–161, 173, 242
Hickson, Paul, 106–108, 169
Hinkley, Sasha, 257
HMS (Humason, Mayall, and Sandage catalog of redshifts), 28–30
Ho, Luis, 178–179
Hoessel, John, 34
Hooker telescope. *See* Mount Wilson Observatory: 100-inch (Hooker) telescope
Hoyle, Fred, 59, 61, 65, 75, 94. *See also* B²FH
HR 8799, exoplanetary system, 254, 256–258, 259

Hubble, Edwin, xi, 6, 7, 14, 16, 21, 23–24, 27, 28, 34, 36, 95–96, 119, 164, 165, 197
   and expansion of the universe, 6, 21, 22, 28, 30, 33, 43, 211
   and redshift-distance relation, 21, 33, 43
Hubble Atlas of Galaxies, 95, 100, 163
Hubble constant (Hubble-Lemaître constant), 22, 29, 30–31, 55
Hubble diagram, 28, 30, 32, 33, 38, 39, 42, 43, 131
   and expansion history of the universe, 30, 39, 43
Hubble eXtreme Deep Field, 2
Hubble Space Telescope (HST), ix, xi, 2, 43, 55, 122, 138, 140, 149, 165, 172, 174, 180, 204, 234, 244, 247, 264–265, 271
   Wide Field and Planetary Camera, 136, 198
Humason, Milton, 28, 91, 198. *See also* HMS
hydrogen bomb, 60–61

Icarus, asteroid, 227
Illustris Project, 196
Image-tube Photon Counting System (IPCS), 186–189
infrared astronomy, 17, 55, 124, 130, 140–156, 170–173, 179, 191, 242, 246–252
   early work in, ix, 17, 140–142
   luminous infrared galaxies, 152–153
   and planetary science, 215–222, 225–227, 238, 252

   and search for exoplanets, 254–257
   and supernovae, 41–42
   technology of, xiii, 17, 141, 215–216, 225
   ultraluminous infrared galaxies, 153–156
Infrared Astronomy Satellite (IRAS), 150–151, 153, 154, 171
Infrared Processing and Analysis Center (IPAC), 170, 250, 267
Ingersoll, Andrew, 221
International Astronomical Union (IAU), x, 24–25, 26, 235
IPCS. *See* Image-tube Photon Counting System
IRC+10216, carbon star, 145–146

Jansky, Karl, 113
Jet Propulsion Laboratory (JPL), 150, 198, 215, 216, 218, 229, 256
Jewitt, David, 228
Jupiter, 6, 252, 259
   atmosphere, 17, 216–222, 226, 247
   and comets, 223–227, 236, 259
   compared to exoplanets, 253, 255, 259
   compared to smallest stars, 242, 247–248
   Galileo probe visits, 221–222
   Great Red Spot, 218, 220, 221
   Voyager 1 visits, 218–221

Keck Observatory, 43, 192, 208, 209, 234, 250, 254, 266
Kellogg Radiation Laboratory, 59

Kepler, space telescope, 253
Kirkpatrick, Davy, 249, 250
Kirshner, Robert, 40–41, 42
Kitt Peak National Observatory, 104, 166, 174, 186, 197–198, 200, 267
Kowal, Charles, 38–40, 42, 235–236
Kraft, Robert, 27
Kristian, Jerome, 132
Kuiper, Gerard, 232
Kuiper belt, 214, 228–238, 254, 259
Kulkarni, Shri, 245, 246, 256
Kumar, Shiv, 242, 250

Larson, Richard, 144
Las Campanas Observatory, 62, 245
Laser Interferometer Gravitational-Wave Observatory (LIGO), 61, 267
Leavitt, Henrietta Swan, 21, 24
Leighton, Robert, 141–142, 143, 147
Levy, David, 223. *See also* Shoemaker-Levy 9
Li, Weidong, 264–265
luminous infrared galaxies (LIRGs), 152–153
Luu, Jane, 228
Lyman, Theodore, 184
Lyman-α absorption/emission line, 136, 137, 184–186, 189–190, 192, 209–211, 212
Lyman-α blobs, 209–211
Lyman-α spike, 184. *See also* Gunn-Peterson trough
Lyman break, 206–209, 211
Lyman-limit absorption systems, 204–206
Lynden-Bell, Donald, 73–77, 99, 197. *See also* ELS
quasars and dormant black holes, 153
Lynds, Roger, 104, 166, 167, 169, 186

Main sequence. *See* stellar main sequence
Makemake, dwarf planet, 234–235
Markarian 231, galaxy, 112
Mason, Max, 9
Matthews, Keith, 149, 218, 224–225, 246–247
Matthews, Thomas A., 118, 119–122, 127, 129, 159
Mayall, Nicholas, 28. *See also* HMS
Mayor, Michel, 252
Meier, David, 206
Merrill, Paul W., 56, 59, 60, 71
discovery of technetium in stars, 56
Messier 3 (M3), globular cluster, 48–55, 75
Messier 31 (M31), Andromeda galaxy, 6, 7, 16, 23–27, 30, 48, 84, 107, 109, 147–148, 175–176
Messier 51 (M51), Whirlpool galaxy, 90–91, 99–102
Messier 81 (M81), galaxy, 166, 169, 178–179
Messier 82 (M82), galaxy, 166–169, 197
Messier 87 (M87), galaxy and radio/X-ray source, 15, 35, 174–175, 176
Messier 101 (M101), Pinwheel galaxy, 30, 263–264
metals, in stars, 71. *See also* crystal stars; ultraviolet color excess
astronomical meaning of, 71–72
in brown dwarfs, 249, 252

and galactic evolution, 187, 189–190, 193, 204, 209, 211
range in abundances of, 64, 71–80, 82, 160–162, 266
Metzger, Mark, 192
Milky Way galaxy, 6, 21, 23, 27, 48–49, 52, 55, 58, 63–64, 67, 109, 261
center of, 17, 113, 140–141, 147–149
formation of, 70–77, 78–82, 82–84, 161, 184, 197
future disruption of, 109
radio sources in, 113–114, 120
Millennium Run, 196
Miller, William C., 167
Minkowski, Rudolph, 31, 95, 121
on radio sources, 93, 113–119, 121, 126, 133, 198, 209
on supernova light curves, 36–37, 38, 40, 42, 266
Mira, star, 56–57
Mount Wilson Observatory, ix, xi, 3, 5, 6–12, 16, 17, 22–23, 25, 36, 56–57, 62, 71, 141, 183, 187, 271
60-inch telescope, ix, 5, 6, 11, 90, 143, 147
100-inch (Hooker) telescope, ix, 6–7, 11, 14, 23, 24, 28, 75, 90, 124
"moving groups" of stars, 82
multichannel spectrometer, 40, 78
multiple stepped exposures, 165
multislit spectroscopy, 201
Münch, Guido, 143–144
Murray, Bruce, 215–217

Nakajima, Tadashi, 245, 250, 256
National Aeronautics and Space Administration (NASA), 149–150, 227, 242, 243, 251, 253, 263
National Energy Research Scientific Computing Center (NERSC), 261, 263
National Geographic Society, 14
National Geographic Society–Palomar Observatory Sky Survey (NGS-POSS). *See* Palomar Observatory Sky Survey
Naval Air Weapons Station (China Lake, CA), 17, 215
Near-Earth Asteroid Tracking (NEAT), survey, 229
Neugebauer, Gerry, 141–142, 144, 145, 147–149, 160, 223. *See also* Becklin-Neugebauer Object
neutron stars, 36, 67, 265
merging of, 61–62, 82, 260
NGC 1068, galaxy, 149, 175
NGC 1569, galaxy, 164–166
NGC 5128 (Cen A), galaxy, 92–94
NGC 5907, galaxy, 70
NGC 7252 ("Atoms for Peace" galaxy), 109
Nicholson, Philip D., 224, 225
Norma Cluster of galaxies, 203
Nugent, Peter, 263, 265

Oemler, Augustus, 197–198, 200, 202
Oke, J. Beverly, 34, 40, 41, 124–127, 132–133, 199

Index

as instrument builder, 40, 78, 132
Omega Centauri, globular cluster, 81
Oort, Jan, 232
Oort cloud, 214, 228, 230–232, 235, 236, 238, 259
Oppenheimer, Rebecca (formerly Ben R. Oppenheimer), 8, 245, 246, 248, 250, 255–256, 257, 270
Orion Nebula, 143, 242
Ostriker, Jerry, 241

Pale Blue Dot, 253
Palmer, Henry, 119
Palomar 5, globular cluster, 82–83
Palomar Billion-Year Survey, 245
Palomar-Green catalog, 135, 152
Palomar Observatory
    60-inch telescope, xi, 260
    200-inch (Hale) telescope, ix, xi, 3, 15, 16, 21, 22, 24, 28, 31
        Cassegrain cage and focus, 8, 78, 132, 136, 175, 177, 219, 246
        coudé focus and spectrograph, 56–57, 62–63, 72–73, 187, 188
        design and building of, ix, 7–10
        east arm, 143, 216, 219
        prime-focus cage, 8, 96, 122–124, 198–199
    Monastery, xiii, xiv, 108, 136, 177, 188
    Schmidt camera, 18-inch, xi, 12–14, 16, 22, 36, 87, 90, 91, 134–135, 223, 240, 260, 261
    Schmidt camera, 48-inch, xi, 14–16, 22, 31–32, 87, 95, 97, 113, 115, 227, 228–229, 230, 235, 260, 261–263
Palomar Observatory Sky Survey (POSS or NGS-POSS), 14–16, 32, 88, 95, 97, 99, 106, 121, 145, 150, 153, 185, 188, 192
    digitized Palomar Observatory Sky Survey (DPOSS), 192, 251
    peculiar objects found in, 95
Palomar Transient Factory (PTF), 260, 261–263, 266–267
    data-flow of, 262
    observations of nearby Type Ia supernova, 263–265
Payne-Gaposchkin, Cecilia, 266
Pease, Francis, 7
Peimbert, Manuel, 178
Persson, Eric, 55
Peterson, Bruce, 184–185
Pettini, Max, 207
Phillips, Mark, 42
Pinwheel galaxy. *See* Messier 101
Piro, Anthony, 264
planetary nebulae, 46–47
Planet Nine, 236–238, 267
Planet X, 235–236
Pluto, 214, 229, 230, 234–235, 238, 259
Population I and Population II stars. *See* stellar populations I and II
Porter, Russell W., 123, 271
Prime Focus Universal Extragalactic Instrument (PFUEI), 136–137, 198–200
Project 1640, 256–258, 260
protostars, 143–146
Pskovskii, Yuri, 41–42

Quaoar, dwarf planet, 229, 236
quasars, 16, 31, 111–138
    as cosmological sources, 131–133
    discovery of, 121–122, 124–126
    as drivers of gas outflows in galaxies, 180
    energy distribution compared to a star, 130
    as engines in galaxy nuclei, 138
    origin of term, 129
    spectra of, 125, 186, 189
Queloz, Didier, 252
Quimby, Roger, 266

Rabinowitz, David, 229, 230
radio astronomy, 113–138, 140, 147–149, 150–151, 183–185, 191
    early work in, 17, 113–119
    Palomar Observatory and, 115, 118, 119–120, 122
radio telescopes, 16, 115, 118, 122
ram-pressure stripping, 202–203
rapid neutron capture, 60, 61
Reber, Grote, 113, 115
redshift-distance relation for galaxies, 33, 116
redshift-magnitude relation for galaxies, 28–29, 34, 38–40, 55
Reimers, Dieter, 58
Richstone, Douglas, 176
ring galaxies, 104–105
    discovery by Zwicky, 91
Robolo, Rafael, 243
Rockefeller Foundation, The, xi, 3, 7, 9, 10, 11, 14

Roman, Nancy Grace, 71–73, 75
    discovery of ultraviolet color excess, 71–72
    influence on Greenstein's work, 72
    influence on Sandage's work, 72–73
Roque de los Muchachos Observatory, 206
Rose, Wickliff, 7–8, 9
Rubin, Vera, 108–109

Sagan, Carl, 253
Sagittarius A, radio source, 113, 147–149
Salpeter, Edwin, 185
Sandage, Allan, 30, 44, 48, 51, 54, 63, 72, 77, 83, 94, 95–96, 100, 108, 148, 163, 164–165, 166–167. *See also* ELS; HMS; *Hubble Atlas of Galaxies*
    composite color-magnitude diagram, 52–55, 63
    and cosmic distance scale, 28–31, 32–33, 34, 41
    and galactic archaeology, 72–77, 80, 94, 131, 197
    and Hubble constant, 29–30
    on quasars, 120–122, 126–131, 134–136, 159, 206

    search for stellar main sequence of Messier 3, 48–52
Sanders, David, 150–156
Sargent, Wallace L. W., 159–162, 164, 174–175, 176–179, 187, 188, 189, 204–206
Schmidt, Bernhard Voldemar, 11–12

Index

Schmidt, Maarten, 31, 121, 123, 128, 130, 133, 150–151, 152–153, 185–186, 206
    distribution of quasars, 133–135, 138, 173
    Lyman-α spike, 183–185
    in prime-focus cage, 123
    quasar 3C273, 122, 124–128
    and radio sources, 121
Schneider, Donald, 136–137, 205, 206
Schwarzschild, Martin, 51–52
Schwarzschild radius, 175
scientific and technological revolutions, 3, 4, 6
    Hale's leadership in, 4–6
    introduction of charge-coupled devices (CCDs), 198, 229
    rise of infrared astronomy, 17, 140–142, 215
    rise of radio astronomy, 16–17, 113–115, 117
Searle, Leonard, 160–162
    compact dwarf galaxies, 160–162
Searle-Zinn model of galaxy formation, 78–82, 82. *See also* ELS
Second Reference Catalogue of Bright Galaxies (de Vaucouleurs et al.), 163
Sedna, dwarf planet, 230, 235, 236, 238

Seyfert galaxies, 132, 172, 175, 176, 177, 179, 200, 202
Shapley, Harlow, 6, 18, 21, 25
Sheppard, Scott, 236
shock fronts, 145, 158, 170
Shoemaker, Carolyn, 223
Shoemaker, Eugene, 223, 226, 228
Shoemaker-Levy 9, comet, 223–228
Shortridge, Keith, 188
Sirius B, white dwarf star, 66
Sloan Digital Sky Survey (SDSS), 82, 269
slow neutron capture, 60, 61
SMBH. *See* black holes, supermassive
Smith, Graham, 115
Soifer, Thomas, 150–156
solar system, 213–238. *See also individual objects by name*
    dwarf planets in, 229–236, 238
    Jupiter's dominance of, 223
    Kuiper belt, 228, 232
    Oort cloud, 228, 232
    Planet Nine, 236–237
    Planet X, 235–236
    Pluto as dwarf planet, 234–235
    rubble in, 229
    scale of, 231
    schematic map of, 214
"spherical bastards," xiii, 14. *See also* Zwicky, Fritz
Spitzer, Lyman, Jr., 93, 94, 202
Spitzer Space Telescope, 171, 179, 259
standard candles, 21, 30, 31, 32, 35, 55
    Cepheid variable stars as, 21, 24
    RR Lyrae variable stars as, 23, 48
    Type Ia supernovae as, 38, 39–42, 263, 265
starbursts, 151, 153, 155–166, 168–171, 172–173, 180, 201, 204, 211–212
Steidel, Charles, 192, 204–210
stellar main sequence, 48–55, 65, 72–74, 120, 201, 249, 254, 264–265
    and stellar evolution, 50, 51–54, 242
    turnoff point from, 50–52, 55

stellar populations I and II, 23–24, 27, 161
stellar streams, 70, 82–84, 88, 91, 94, 96, 99, 101–102, 110
Stephan's Quintet, 158, 169–171
subdwarf stars, 71–75
sun, 6, 21, 65, 67, 74, 113, 141, 265. *See also* solar system
   death of, 65
   evolution of, 65, 184, 192, 214
supermassive black holes. *See* black holes, supermassive
supernovae, 36–43
   light curves of, 36, 37, 41–42
   production of chemical elements in, 55–56, 60–62, 77
   SN 1937C, 61
   SN 2011fe, 263, 264–265
   as sources of radioactive material, 61
   spectra of, 36, 37, 40
   as standard candles, 38, 39–42, 263, 265
   superluminous, 266–267
   Type I, 36–42, 266
   Type Ia, 42, 67, 261, 263–265, 267
   Type Ib, 42
   Type II, 36, 42, 266
superwinds. *See* winds, galactic
Swope, Henrietta, 25–27
   "Swope slope," 27
Swope telescope, Las Campanas Observatory, 61

Tammann, Gustav, 30–31, 41
Tarter, Jill, 243
technetium, in stars, 56, 59, 60, 61
Terrile, Richard, 218–222
Theys, John, 104
Third Cambridge Catalogue of Radio Sources (3C), 114, 119, 131, 133, 183
   3C 9, 133, 183–185
   3C 48, 120–121, 126, 127, 132–133, 154
   3C 273, 121–122, 124, 125, 126, 127, 128, 132, 154
   3C 295, 118–119, 126, 133, 198, 200–201, 209
tidal interactions, 88, 108, 155, 159, 163–164, 171
Tinsley, Beatrice, 34, 163–164, 169, 197
Toomre, Alar and Juri, 99–106, 108, 155, 163, 169
   numerical modeling of gravitational interactions, 99–106, 110, 197
   role of collisions in galaxy evolution, 102–106
   on tidal interactions, 164
Torres-Peimbert, Silvia, 178
transients, 260–263
Trapezium star cluster, 143, 145
Trimble, Virginia, 66
triple-alpha process, 59
Trujillo, Chad, 229, 230, 236
Tsuji, Takashi, 248–249, 256
Tuton, Gary, 199
Two-Micron Survey, 141
Tytler, David, 189

Ultraluminous infrared galaxies (ULIRGs), 153–156
ultraviolet color excess, 72–75, 77

Index

Vaucouleurs, Gérard de, 163
Vela, system of satellites, 191
velocity-distance relation. *See* Hubble diagram
Visvanathan, Natarajan, 167
Vorontsov-Velyaminov, Boris, 16, 95, 106
Voyager 1, NASA mission, 219–221, 253

Wallerstein, George, 63–64
warm-hot intergalactic medium (WHIM), 182, 190
Westphal, James, 143, 174, 198, 215, 218–222, 244, 252
   hotspots on brown dwarf stars, 252
   hotspots on Jupiter, 218–219, 221–222
   introduction of CCDs to Palomar, 198
Whirlpool galaxy. *See* Messier 51
white dwarf stars, 47, 50, 56, 60, 65–67, 129, 145, 264–265, 267
   carbon-oxygen, 67, 264
   crystallized interiors, 67
   electron degeneracy in, 65
   merging white dwarfs, 267
Wildey, Robert, 216
winds, galactic, 166–169, 211–212
winds, stellar, 47, 50, 55, 60, 77, 112, 145, 146, 153, 161–162, 165

Xu, Kevin Cong, 170–171

Young, Peter, 174–175, 189

Zimmerman, Barbara, 199
Zinn, Robert, 78–82
Zwicky, Fritz, xiii, 11–14, 36, 38, 62, 86–92, 94, 99–100, 104, 105, 107, 109–110, 135, 159–160, 241
   discovery of Cartwheel galaxy, 105
   on gravitational galaxy interactions and tides, 86–92
   inference of dark matter, 241
   as "spherical genius," 11–14
   supernova searches, 36–37, 43, 58, 260, 266
   surveyor and cataloger, 11, 13, 32, 87, 106, 153, 160, 162
Zwicky, Margrit, 153
Zwicky Transient Facility (ZTF), 266–267